CAROTENOIDS

Edited by **Dragan J. Cvetković**
and **Goran S. Nikolić**

Carotenoids
http://dx.doi.org/10.5772/65523
Edited by Dragan J. Cvetković and Goran S. Nikolić

Contributors

Rashidi Othman, Oguz Merhan, Hernan Ceballos, Fabrice Davrieux, Elise Talsma, Meike Andersson, Diana Barraza-Jimenez, Manuel Flores-Hidalgo, Miguel Escobedo-Bretado, Daniel Glossman-Mitnik, Jesus Monzon-Bensojo, Francisco Torres-Rivas, Lucia Maria Jaeger De Carvalho, Gisela Maria Dellamora Ortiz, Jose Luiz Viana Carvalho, Lara Smirdele, Flavio De Souza Neves Cardoso, Volker F. Wendisch, Nadja Alina Henke, Petra Peters-Wendisch, Sara Mendiara, Luis Perissinotti, Kazuo Yamagata, Cecilia Faraloni, Giuseppe Torzillo, Ligia Cardoso, Karen Kanno, Susan G. Karp, Francielo Vendruscolo, Liliana Isabel Zoz, Júlio César Carvalho, Kazimierz Strzalka, Malgorzata Jemiola-Rzeminska, Paulina Kuczynska, Jerilyn Timlin, Maria Shumskaya, Aaron Collins, Eleanore Wurtzel, Thomas Beechem Iii

Published by InTech
Janeza Trdine 9, 51000 Rijeka, Croatia

Publishing Process Manager Mirena Calmic
Technical Editor SPi Global
Cover InTech Design team

First published June, 2017
Printed in Croatia
Legal deposit, Croatia: National and University Library in Zagreb

Additional hard copies can be obtained from orders@intechopen.com

Carotenoids, Edited by Dragan J. Cvetković and Goran S. Nikolić
p. cm.
Print ISBN 978-953-51-3211-0
Online ISBN 978-953-51-3212-7

For EU product safety concerns:
IN TECH d.o.o., Prolaz Marije Krucifikse Kozulić 3, 51000 Rijeka, Croatia,
info@intechopen.com or visit our website at intechopen.com.

Meet the editors

Dr. Dragan Cvetković was born on 26 June, 1977, in Leskovac. He finished his elementary and high school education in Lebane, while he completed his studies at the Faculty of Technology in Leskovac in the year 2002 and finished his PhD thesis in the year 2012 at the Faculty of Technology in Leskovac. Dragan Cvetković participated in the realization of numerous projects funded by the Ministry of Science, Republic of Serbia. He was engaged in the project 'Folding and Stability of Phycobilisome Proteins' at the Institute of Biology and Technology of Saclay, France. He also participated in the realization of the project entitled 'Contribution of Chemical Quenching of Singlet Oxygen to Pro- and Antioxidant Activity of Carotenoids', funded by the Polish Ministry of Science. He was elected as a teaching assistant in the year 2008 for physical chemistry, colloid chemistry and instrumental analysis, and in the year 2012, he was elected as an assistant professor for a physico-chemical group of subjects at the Faculty of Technology in Leskovac.

Dr. Goran Nikolić was born in Knez Selo (Niš, Serbia) on 1 November 1966. He received his PhD degree in Chemical Engineering (2001) from the University of Niš. Currently, he is a full professor at the same university, for pharmaceutical-cosmetic engineering group of subjects at the Faculty of Technology in Leskovac. His research activities are development of pharmaceutical products and quality control of drugs. His competences are experience in project management and managing academic institutions at different levels. He is a member of several national projects in the technological development area and is a member of numerous TEMPUS Joint European projects of sustainable technologies, environmental application and management courses. He has authored more than 300 scientific papers, numerous technological solutions for pharmaceutical industry, national monographs, international patents, university textbooks and invitation lectures. He is an editor of two international monographs on FTIR spectroscopy (InTech Open Access Publisher).

Contents

Preface

Carotenoids are one of the most widespread pigment groups distributed in nature; more than 700 natural carotenoids have been described so far. This group of pigments is known for versatile roles they play in living organisms; however, their most pivotal function is involvement in scavenging of reactive oxygen species and photoprotection.

Within the chapters in this book, different activities (Section I) and biotechnological applications (Section II) of carotenoids are discussed.

An important photoprotective mechanism, referred to as the xanthophyll cycle, and the mechanisms of less-known types of the mentioned cycle are described; the current knowledge on the pigments engaged in the xanthophyll cycles operating in various organisms is also summarized. Electronic properties of six carotenoids such as energy in frontier orbitals and the first molecular orbitals to work in the UV-Vis absorption spectroscopy are also analyzed. Electronic structure methodologies were used within the frame of density functional theory using different theoretical methods. Results for the main absorption peak are in agreement with experimental results, and possible use in energy generation systems is discussed. In addition to theoretical studies, the application of Raman spectroscopy, providing detailed information about the molecular structure of carotenoids, is examined in the light of a powerful method for mapping carotenoid abundance in cells and tissues. In many applications, this technique is compatible with living organisms, providing highly specific molecular structure information in intact cells and tissues with subcellular spatial resolution.

Antioxidant activities of carotenoids are discussed through several chapters. Reaction of β-carotene with nitrogen dioxide and nitric oxide, both in pure dioxane and dioxane/water solvent, is investigated by EPR and UV-Vis spectroscopy. An appropriate discussion on the importance of carotenoid compounds and their reactions in biological media, as well as the role and the possible reactions of nitroxide intermediates, is evaluated. The biosynthesis of carotenoids as a response on stress factors in order to investigate variables that regulate carotenoid accumulation through manipulation of drought stress, light intensity and nutrient strength is analyzed. Finally, an overview of carotenoid effects in cardiovascular disease risk reduction, their unique biological activities in staving off atherosclerosis and endothelial dysfunction mitigation is given.

The second part of the book deals with biotechnological applications of carotenoids isolated from different sources. Carotenoids as natural pigments with important biological activities, such as antioxidant and provitamin A activity, have a great potential in the food, feed and pharmaceutical industries. They can be either extracted from plants and algae or synthesized by various microorganisms, including bacteria, yeasts, filamentous fungi and microalgae. Microbial production, including the ability of microorganisms to use a wide variety of low-cost substrates, the better control of cultivation and the minimized production time, has a number of advantages in carotenoid production. Under stress conditions such as high light exposure, nutrient starvation, change in oxygen partial pressure and high or low temperatures, microalgal metabolism is altered, and photosynthetic activity may be reduced. In these conditions, photosynthetic electron transport is reduced, and the increase of intracellular reduction level may be associated to the formation of free radicals and species containing singlet oxygen. In order to prevent damage from photo-oxidation, microalgae are able to adopt strategies to contrast these dangerous oxidant molecules. One of the most active mechanisms is to synthesize large amount of carotenoids, which can act as antioxidants. Among microalgae, *Haematococcus, Chlamydomonas, Chlorella* and *Dunaliella,* as well as diatoms and dinoflagellates, such as *Phaeodactylum* and *Isochrysis,* are able to synthesize large amount of carotenoids. The main primary carotenoids usually found are neoxanthin, violaxanthin, lutein and b-carotene. To preserve cells from oxidative damage, their production may be increased, while other carotenoids may be synthesized de novo. A yellow-pigmented soil bacterium *Corynebacterium glutamicum* is also used as a workhorse of industrial biotechnology for more than 60 years. Carotenoids, the colourful representatives of terpenoids, are high-value compounds whose bio-based productions are on the rise. Since *C. glutamicum* is a natural producer of the rare C50 carotenoid decaprenoxanthin, this organism is well suited to establish terpenoid-overproducing platform strains with the help of metabolic engineering strategies. The carotenogenic background of *C. glutamicum* and the metabolic engineering strategies for the generation of carotenoid-overproducing strains are depicted. In the same time, vitamin A deficiency is a preventable tragedy that affects millions of people, particularly in sub-Saharan Africa. During the last two decades, significant efforts have been made to identify sources of germplasm with high provitamin A carotenoids in an attempt to use them in development of cultivars with satisfactory nutritional characteristics and good agronomic performance. The cassava root is especially examined since a large proportion of people in sub-Saharan Africa relies on diets based on cassava as a source of calories. Furthermore, the presence of provitamin A carotenoids, mainly the β-carotene in pumpkins, yellow sweet potato and yellow sweet and bitter cassava, is also reported.

Prof. Dragan J. Cvetković and Prof. Goran S. Nikolić
University of Niš
Faculty of Technology, Leskovac Serbia

Section 1

Carotenoids: Characterization and Activity

Chapter 1

Characterisation of Carotenoids Involved in the Xanthophyll Cycle

Paulina Kuczynska,
Malgorzata Jemiola-Rzeminska and
Kazimierz Strzalka

Additional information is available at the end of the chapter

http://dx.doi.org/10.5772/67786

Abstract

Carotenoids are known for versatile roles they play in living organisms; however, their most pivotal function is involvement in scavenging reactive oxygen species (ROS) and photoprotection. In plant kingdom, an important photoprotective mechanism, referred to as the xanthophyll cycle, has been developed by photosynthetic organism to avoid excess light that might lead to photoinhibition and inactivation of photosystems and induce the formation of reactive oxygen species (ROS), resulting in photodamage and long-term changes in the cells caused by oxidative stress. Apart from high-light driven enzymatic conversion of violaxanthin (Viola) to zeaxanthin (Zea) that occurs mostly in higher plants, mosses and lichens, other less known types of the xanthophyll cycle have been hitherto described. The work is aimed at summarising the current knowledge on the pigments engaged in the xanthophyll cycles operating in various organisms.

Keywords: carotenoids, chromatography, diadinoxanthin, diatoms, diatoxanthin, *Phaeodactylum tricornutum*, xanthophyll cycle

1. Introduction

Carotenoids constitute a large group of pigments with over 700 compounds [1]. They comprise of carotenes and their oxygenated derivatives, xanthophylls. Carotenes are polyunsaturated hydrocarbons with 40 carbon atoms, while xanthophylls contain oxygen atoms, most frequently as hydroxyl and epoxide groups, which increase their polarity. Both groups of carotenoids act as accessory light-harvesting pigments or as quenchers of singlet oxygen and chlorophyll triplet states to provide protection against photooxidative damage [2]. The main photoprotective mechanism occurring in photosynthetic organisms is the xanthophyll cycle—a process of enzymatic reactions of epoxidation and de-epoxidation of xanthophylls [3]. These cyclic conversions can proceed between several pigments including violaxanthin (Viola), antheraxanthin (Anth), zeaxanthin (Zea), diadinoxanthin (Diadino), diatoxanthin (Diato), lutein (Lut), lutein-epoxide (LutE) and oxidised but not epoxidised siphonaxanthin (Siph). Therefore, five types of xanthophyll cycles and additional non-specific cycle have been described [4] (**Figure 1**). The common factor in all of them is conversion of epoxidised xanthophylls to their de-epoxidised forms under strong light to dissipate of excess energy and epoxidation of de-epoxidised xanthophylls in low light or dark [5].

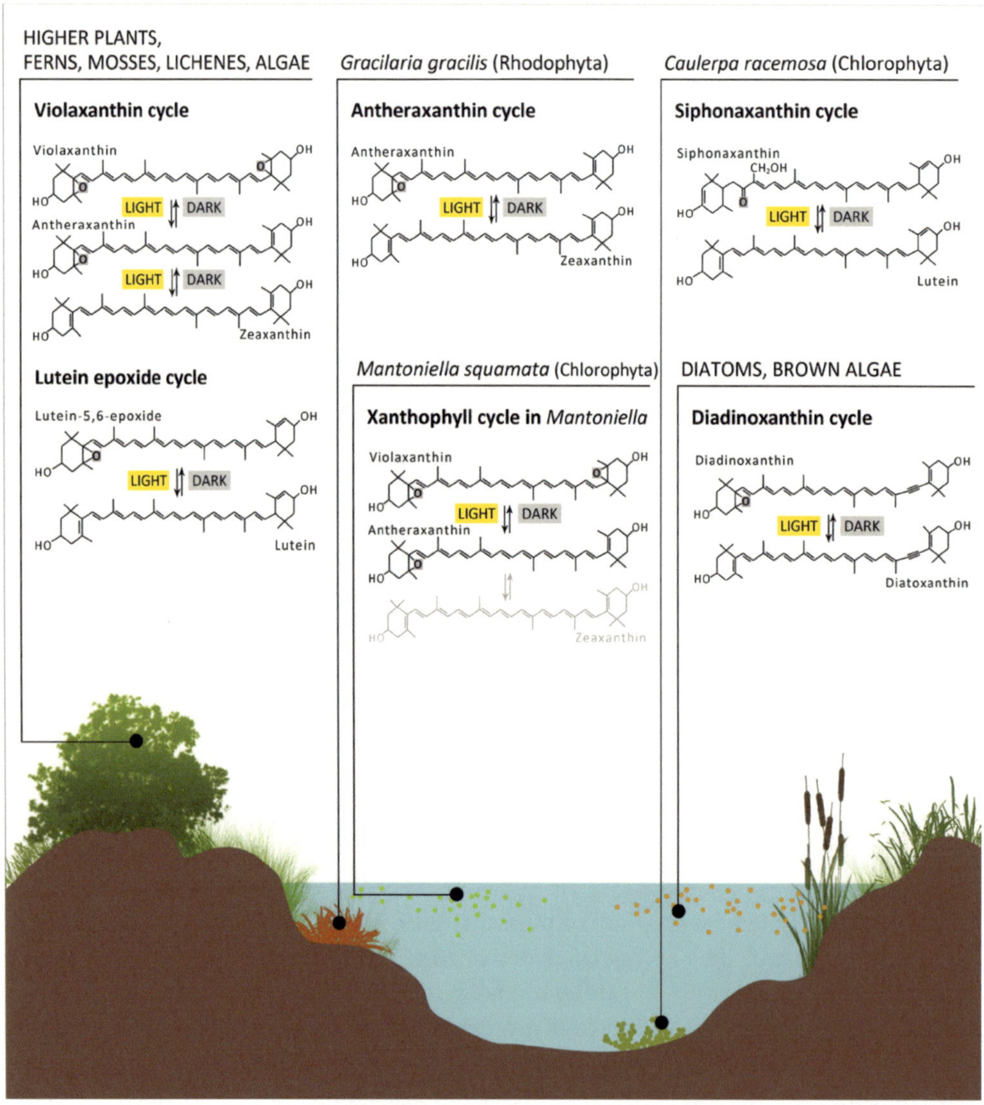

Figure 1. Xanthophyll cycles in photosynthetic organisms.

2. Carotenoids in the xanthophyll cycle

2.1. Violaxanthin, antheraxanthin and zeaxanthin

Viola, Anth and Zea are engaged in the most common xanthophyll cycle, referred to as the violaxanthin cycle (VAZ cycle), see **Figure 1**. Di-epoxy Viola is de-epoxidised to epoxy-free Zea in two-step reaction catalysed by Viola de-epoxidase (VDE) with mono-epoxy Anth as an intermediate product. In absence of photosynthesis, VDE is localised in the thylakoid lumen as inactive monomer; however, it undergoes dimerization and binds to the membrane in an acidic pH caused by the light-driven transmembrane proton gradient [6–8]. Additionally, ascorbate as a donor of protons and monogalactosyldiacylglycerol (MGDG) as lipid-forming inverted hexagonal structures are essential for Viola de-epoxidation [9, 10]. In reverse reactions of the VAZ cycle, product of de-epoxidation—Zea—is epoxidised by Zea epoxidase (ZEP) to Viola, also via Anth. The reaction is observed in low light and in darkness due to lack of VDE activity in such conditions, but in higher plants, it can also proceed in high light [11, 12]. ZEP, localized in chloroplast stroma, is active in neutral pH and requires nicotinamide adenine dinucleotide phosphate (NADPH), flavin adenine dinucleotide (FAD), and molecular oxygen as co-substrates to epoxidise rings in Viola and then in Anth [11, 13].

The VAZ cycle occurs in higher plants, ferns, mosses, lichens and some groups of algae [3, 4]. However, in few species of algae, two specific xanthophyll cycles with Viola, Anth and Zea have been observed (**Figure 1**). In *Mantoniella squamata* (Chlorophyta), Viola is converted mainly to Anth, which is rapidly epoxidised back to Viola, and Zea occurs in low amount in these cells. It is a result of reduced affinity of VDE to Anth [14]. Second modification in the VAZ cycle has been described in two Rhodophyta species *Gracilaria gracilis* and *Gracilaria multipartita* in which Viola does not occur so de-epoxidation and epoxidation proceed only between Anth and Zea [15].

2.2. Diadinoxanthin and diatoxanthin

In several algal groups including diatoms, phaeophytes, dinophytes and haptophytes, mainly Diadino and Diato are involved in the xanthophyll cycle therefore named diadinoxanthin cycle (DD cycle). In addition, also the VAZ cycle can be observed in these organisms during strong light stress [16, 17], see **Figure 1**. In DD cycle, mono-epoxy Diadino is converted to epoxy-free Diato by a Diadino de-epoxidase (DDE or VDE), and the reverse reaction is catalysed by a Diato epoxidase (DEP or ZEP). Both enzymes have comparable properties and are able to convert Diadino/Diato as well as Viola/Anth/Zea [17]. De-epoxidation occurs in high light, in decreased pH in thylakoid lumen and in the presence of ascorbate, however, at lower concentration than that in plants [18, 19]. Unlike plants, in diatoms, epoxidation does not occur in high light since the proton gradient between thylakoid lumen and chloroplast stroma inhibits this reaction [20, 21]. It was reported that the rates of Diadino de-epoxidation and Diato epoxidation are several times higher than the conversions of Viola and Zea in plants and green algae [20].

2.3. Lutein, lutein epoxide and siphonaxanthin

Lut is an epoxy-free xanthophyll bound to antenna proteins and is essential for their stability in higher plants, while LutE occurs in significant amount in some species only [22], see **Figure 1**. Although the presence of LutE does not mean its involvement in cyclic conversions with Lut, the fully operative or truncated LutE cycle has been reported in several plant species [4]. In the first case, the initial LutE pool is fully recovered in the dark, which is not observed in truncated cycle. Both reactions of the LutE cycle are catalysed by VDE and ZEP, which are also engaged in the VAZ cycle but their rates are 2- or 3-fold lower, which might be a result of decreased affinity of enzymes to Lut and LutE or stronger binding of these pigments to antenna proteins [4].

Another xanthophyll cycle is operating in green algae *Caulerpa racemosa* in which interconversions between Siph and Lut have been reported [23]. During illumination, Siph is converted to Lut, and the reverse reaction proceeds in low light. Despite the mechanism of the Siph cycle is still unknown, this suggests a photoprotective role [22].

3. *In vitro* assays of the xanthophyll cycle

Studies on de-epoxidation and epoxidation of pigments involved in the xanthophyll cycle, the mechanism and conditions of these reactions, enzyme properties and factors regulating their efficiencies are usually performed *in vivo* by treatment of the organisms studied with stress or genetic modifications. Such experiments allow to observe holistic effects in natural or semi-natural conditions but to analyse specific parameters of a single reaction, it is more convenient to perform them in fully controlled system.

Among two reaction types in the xanthophyll cycles, de-epoxidation of Viola has been extensively studied, and *in vitro* assay was developed [24]. The system comprises phosphatidylcholine liposomes with MGDG and Viola suspended in sodium citrate buffer with sodium ascorbate and VDE isolated from wheat. It was concluded that MGDG is an essential component of lipid membrane, which allows to bind VDE to the membrane which is necessary for its activity. Additionally, de-epoxidation of Viola to Anth seemed to be more sensitive to MGDG concentration than the second step of the reaction. Although this assay has been tested with Viola as a substrate, it is highly probable that Diadino can be also used for such assay. Considerable progress in these studies is the use of purified recombinant enzymes preferably from several species which allow for comparative analysis of their properties.

Epoxidation of Zea was investigated in semi-defined system [25]. Thylakoids of *npq1* mutant of *Arabidopsis thaliana* were the source of an active ZEP suspended in Hepes buffer with sorbitol, $MgCl_2$ and ethylenediaminetetraacetic acid (EDTA). Zea was mixed with MGDG and incorporated into thylakoids by sonication, and also sodium ascorbate, FAD and NADPH were added. During 2 hours, the amount of Zea was reduced by 38%. However, isolated thylakoids contain not only an active enzyme but also additional compounds that could play an essential role in Zea epoxidation. Therefore, such semi-defined system may not be applicable to study of purified enzyme activity.

In vitro assay of Zea epoxidation has also been reported [26] using recombinant *Capsicum annuum* β-cyclohexenyl epoxidase isolated from *Escherichia coli*. Reaction was carried out in phosphate buffer with FAD, NADPH, Zea, MGDG, digalactosyldiacylglycerol (DGDG), ferredoxin, ferredoxin: $NADP^+$ oxidoreductase and β-cyclohexenyl epoxidase. It has been reported that epoxidase is able to accept NADP only via reduced ferredoxin activity in the presence of NADPH, ferredoxin oxidoreductase and ferredoxin, and in these conditions, a significant conversion of Zea into Viola was observed.

4. Production of the xanthophyll cycle carotenoids

The production of carotenoids with well-known beneficial effects is of great importance for various industries including food, cosmetology, pharmacy and medicine and can be performed both by extraction from plans, algae, fungi, yeast and bacteria or through chemical synthesis [27, 28]. Both methods have some advantages and disadvantages but the choice is usually dependent on availability of extraction or synthesis procedure [29]. Due to many technologies of chemical synthesis have been developed and the cost of this production often is relatively low, the majority of carotenoids is obtained chemically. However, a consequence of that way is usually the production of stereoisomers mixture with reduced biological activity or even having side effects. Such disadvantages are not the case of natural pigments extraction; however, difficulties with yield and separation efficiency are present. Chemical and physical properties of various carotenoids are similar which results in limited separation capacity.

Most of the xanthophyll cycle carotenoids such as Anth, Zea, Viola, Lut and LutE are commercially available (data are given by international carotenoid society), but Diadino and Diato production has been developed only recently [30]. The procedure consists of total pigments extraction from marine diatom *Phaeodactylum tricornutum* followed by saponification and pigments partitioning and finally purification by open-column chromatography (**Figure 2**).

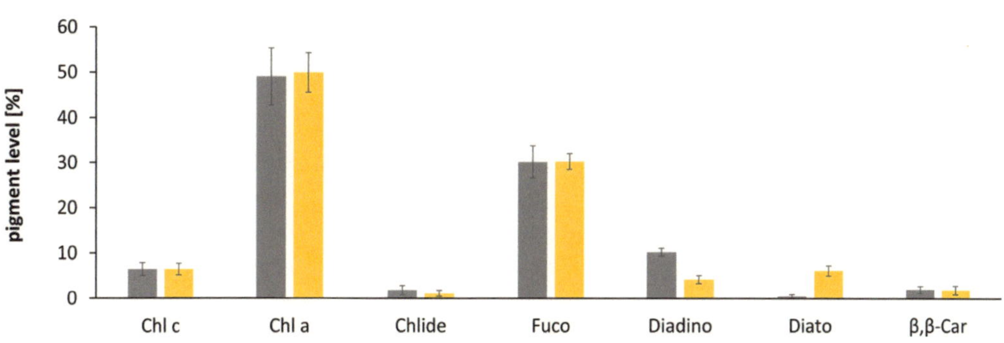

Figure 2. A process of isolation and purification of diadinoxanthin and diatoxanthin illustrating the successive steps comprising: diatoms cultivation under different light conditions, total pigments extraction, carotenoids separation by partitioning and purification of diadinoxanthin and diatoxanthin by open-column chromatography.

4.1. Diatoms as a rich source of natural pigments

Diatoms are microalgae widely distributed in marine and freshwater environment. They contain two types of photosynthetic pigments which are involved in light harvesting (chlorophyll *a*, chlorophyll *c* and fucoxanthin) and photoprotection (β-carotene, diadinoxanthin, diatoxanthin, zeaxanthin, antheraxanthin and violaxanthin). Three of above-mentioned pigments including Fuco, Diadino and Diato occur only in few algal groups; therefore, they might be considered as diatom-specific carotenoids. The quantitative composition of above pigments is dependent on growth conditions.

Pigment content is regulated mostly by light which increases the level of photoprotective carotenoids which are involved in the xanthophyll cycle [31]. Therefore, light stress can be used to produce the highest possible content of Diadino or Diato in the cells.

4.2. Diatoms cultivation

Diatoms *P. tricornutum* Bohlin, strain CCAP 1055/1, were cultivated in f/2-Si medium [32] made with seawater supplemented with inorganic nutrients and vitamin mixture. The temperature of 15°C was proved to be an optimal for increased level of both xanthophylls [30]. Cells were grown under white light of the intensity of 100 μmol photons m^{-2} s^{-1} in a 16/8 hours day/night photoperiod. Cells used to Diadino purification were collected after dark phase, while in the case of Diato, purification cells were illuminated with white light of the intensity of 1250 μmol photons m^{-2} s^{-1} for 2 hours. Several light conditions during diatoms growth have been tested, which resulted in high level of both xanthophylls (even 19% of Diadino or 17% of Diato); however, *cis* isomers of Diadino and Diato and also Zea, Anth and Viola were detected in addition in these conditions [30]. These additional xanthophylls have similar physical and chemical properties, including polarity, which has crucial impact on separation efficiency, and therefore, obtained Diadino and Diato were contaminated by them. Summarising, to obtain pure Diadino and Diato, diatoms should be cultivated under specific light conditions which result in increased biosynthesis of these pigments and simultaneously do not cause induction of the VAZ cycle and *cis* isomers formation. The pigment composition in *P. tricornutum* cultivated in such conditions is given in **Figure 3**, and the levels of Diadino or Diato in these samples were 10 and 6%, respectively.

4.3. Pigment extraction and saponification

An essential aspect of pigment quantification and its collecting for further purification is a step of extraction which requires selection of an appropriate method for a particular purpose. To analyse the amount of pigments in the cells, it is important to extract each of them with the same yield to avoid disproportion between their content. Since pigment composition of each organism varies and includes pigments with different polarity and hence the solubility in organic compounds, the extraction technique should be selected individually. On the other hand, to obtain a particular pigment, special attention should be paid to the extraction yield and to minimising formation of degradation products. In view of these issues, many methods of pigment extraction have been developed. The most applicable methods are solvents that

http://dx.doi.org/10.5772/67786

Figure 3. Pigment composition in *Phaeodactylum tricornutum* cultivated at 15°C under white light of the intensity of 100 μmol photons m^{-2} s^{-1} in a 16/8 hours day/night photoperiod. One day old cells were collected after dark phase (dark) or after illumination with white light of the intensity of 1250 μmol photons m^{-2} s^{-1} for 2 hours (light) [30].

comprise acetone, methanol and water in varying ratios, while the most common homogenisation techniques include grinding, sonication, heating and shaking [33].

The aim of pigment extraction from *P. tricornutum* in described protocol was to recover the highest amount of Diadino and Diato neglecting the efficiency of other pigments extraction. The composition of solvent mixture and the proportion between the solvent volume and the number of cells were found to be essential for the yield of this step. Pigments were extracted from frozen cells, which were earlier harvested by centrifugation of liquid diatoms culture, using a medium composed of methanol, 0.2-M ammonium acetate and ethyl acetate (81:9:10, v/v/v) in a ratio of 10 mL per 2 × 10^9 cells [30].

Saponification and liquid-liquid partitioning are widely practised techniques of carotenoids purification, which allows to remove chlorophylls and lipids—compounds abundantly present in pigment extracts. In general, saponification is performed by adding methanol or ethanol and aqueous solution of KOH in concentration of approx. 5–10% into pigment extract and then incubation of the mixture in darkness for several hours. Numerous modifications of this method have been described [34]. Before performing this step, it should be considered whether the carotenoid is stable in alkali because some of them are not, that is, astaxanthin, fucoxanthin and peridinin [35, 36].

After saponification, chlorophylls and carotenoids are separated through the partitioning with mixture of solvents of varied polarity. Lipophilic carotenoids dissolve in the upper phase, which usually contains petroleum, while chlorophylls dissolve in methanol or ethanol in the lower phase. However, the presence of epoxy, hydroxy and other groups in pigments cause decrease in polarity of chlorophylls, while the polarity of some carotenoids is enhanced [34]. Therefore, the composition of mixture for partitioning should be more complex and specific for the extract.

In the case of Diadino and Diato purification, a saponification was performed through addition of ethanol and 60% aqueous solution of KOH (10:1, v/v) and stirring the mixture in darkness overnight in a cold room. To separate phases following solvents were added consecutively: hexane and diethyl ether (1:1, v/v), extraction petroleum and water (4:1:2, v/v/v). Then, to remove alkali, collected carotenoid fraction was washed three times with one volume of distilled water. The efficiency of a single partitioning was more than 90% so repetitions which resulted in increasing of Fuco level were not advisable. After chlorophylls and carotenoids separation, the level of Diadino or Diato in extract increased to approx. 40 or 20%, respectively, and mixtures contained also β,β-Car, cryptoxanthin and fucoxanthin derivatives. This preliminary purification step was essential for further purification by open-column chromatography.

4.4. Diadinoxanthin and diatoxanthin purification

The final step of Diadino and Diato purification was separation of carotenoids present in the mixture after saponification and partitioning by open-column chromatography. This technique is commonly used in preparative scale and therefore is applicable for industrial use. Development of appropriate chromatographic conditions consists mainly in selection of the solid phase and composition of eluents, but also in the amount of pigment mixture applied onto the column, the volume of eluents, pressure applied to the column.

Since xanthophylls are susceptible to isomerization under acidic conditions, a modified silica gel, which is chemically converted into a basic form, was used. To separate and elute pigments, the mixtures of hexane and acetone were used. Hexane:acetone (90:10, v/v) mixture allows for β,β-Car elution. Second mixture (80:20, v/v) removes cryptoxanthin epoxide from the column. Continued use of this mixture allows to separate and collect Diato. Then, third mixture (70:30, v/v) allows to collect Diadino. To remove Fuco derivatives from the column and to reuse the silica gel, pure acetone was used.

Described procedure allowed to obtain all-*trans* Diadino and all-*trans* Diato of a purity of 99.9%, and the final efficiency was estimated to be 63 and 73% for Diadino and Diato, respectively. Due to the use of popular reagents, simplicity of the procedure, possibility of diatoms cultivation in bioreactors and estimated low costs, the method is widely accessible and might be performed both in analytical and preparative scale.

5. Interaction of the xanthophyll cycle carotenoids with membranes

It has been in 1974 when the hypothesis was formed postulating that carotenoids present in some prokaryotic membranes play the similar role as cholesterol in animal membranes [37]. Since then, numerous studies have been carried out to get a deep insight into the molecular mechanism of carotenoid-membrane interactions. A large body of research has been performed using liposomes as model systems of biological membranes, mostly due to their simplicity, stability, and well-characterised properties.

5.1. Localisation within model membranes

The issue of localisation and orientation of carotenoids in the membrane has been addressed mainly by analysing the position of the maxima in the UV-VIS spectral region of these pigments upon their incorporation into lipids matrix as well as by linear dichroism and X-ray diffraction [38]. Generally, xanthophylls are oriented perpendicularly to the membrane surface with their hydrophilic groups anchored in the two opposite polar regions of the lipid bilayer. Accordingly, to minimalise the energy of the system, zeaxanthin molecule with its two hydroxyl groups located at 3 and 3′ position is thought to span the membrane. Alternatively, all polar groups of a xanthophyll molecule can remain with contact with the same polar headgroup region of the membrane. Such orientation has been proposed for xanthophylls in a conformation *cis* [39]. Interestingly, lutein characterised by the rotational freedom around the 6′–7′ single bond is believed to adopt 2 orthogonal orientations: one roughly vertical and the other horizontal to the lipid bilayer [40, 41].

5.2. The influence on the physical-chemical properties of membranes

The orientation of carotenoids within lipid bilayer enables them to interact with alkyl chains of lipids via van der Waals interactions, which results in modifications of structural and dynamical properties of membranes. Among experimental methods employed to investigate the effect of carotenoids on the properties of biomembranes are electron paramagnetic resonance (EPR), nuclear magnetic resonance (NMR), differential scanning calorimetry (DSC) as well as fluorescence and permeability measurements. An excellent overview of this research is presented in [42]. Considering xanthophylls, Viola, Zea and Lut were reported as modulators of membrane fluidity, increasing it in the ordered phase of the phospholipid bilayer and acting conversely in the liquid crystalline phase [43, 44]. Moreover, in the presence of xanthophylls, an increase of the penetration barrier to molecular oxygen into the hydrophobic region of membrane was observed [45]. In general, polar carotenoids were capable of influencing the thermotropic phase behaviour of phospholipid membranes. As a result, main phase transition ($P_{\beta'} \rightarrow L_{\alpha}$) was shifted to lower temperature in a concentration dependent manner accompanying by a decreased cooperativity and the molar heat capacity of the $P_{\beta'} \rightarrow L_{\alpha}$ transition [46]. Lutein and zeaxanthin affected alkyl chains of lipid bilayers by restriction of molecular motions of both CH_2 and the terminal CH_3 groups [47, 48]. Furthermore, xanthophylls, especially Lut, were responsible for an increase in the thickness of lipid membranes composed of lecithins with myristoyl and palmitoyl moieties [41, 49].

Recently, experiments in atomic force microscope (AFM) showed that the presence of Viola or Zea embedded in the liposomes at a concentration up to 1 mol% does not significantly affect the morphology of the vesicles. However, adhesive forces were 10 times higher for dipalmitoylphosphatidylcholine (DPPC) membranes enriched in Zea, than those observed for an untreated system [50].

Finally, it is worth mentioning that unlike the xanthophylls of the VAZ cycle, diatom-specific carotenoids including Diadino and Diato have hitherto been much less studied in terms of the effect they can exert on the structural and dynamic properties of biomembranes. Interestingly, based on measurements carried out most recently in our laboratory by use of DSC technique and fluorescence anisotropy of 1,6-diphenyl-1,3,5-hexatriene in phospholipid liposomes (manuscript in preparation), it seems that Diadino and Diato affect the lipid bilayer much stronger than Viola and Zea.

6. Conclusion

Along with the characterisation of physiological roles, various carotenoids play being involved in the xanthophyll cycle, *in vitro* assays of both epoxidation and de-epoxidation reactions enable to study the molecular mechanism of the cycle. Moreover, it is interesting to address the question of effect carotenoids have on biological membranes, as it helps to put some light on their antioxidant activity. This in turn seems to play a key role in terms of nutrition and human health. In view of this, the issue of isolation and purification of diatom-specific xanthophylls, yet less described in literature seems to be of great importance. Recently, developed method of all-*trans* diadinoxanthin (Diadino) and all-*trans* diatoxanthin (Diato) purification from *P. tricornutum* comprises four-step procedure and is dedicated to both analytical and preparative scale.

Particular attention is paid to natural carotenoids that apart from the photoprotective function in photosynthetic organism have been recognised as exhibiting beneficial activities for humans and animals and used for commercial and industrial applications. Diatoms seem to be a promising source of unique bioactive compounds, with diadinoxanthin and diatoxanthin as representatives of the xanthophyll cycle pigments. Given that until now neither Diadino nor Diato were commercially available in amounts greater than those used as standards in high-performance liquid chromatography (HPLC), the efficiency of the described purification procedure reaching up to 73% makes the method economically feasible.

Acknowledgements

The Jagiellonian University is a partner of the Leading National Research Center (KNOW) supported by the Ministry of Science and Higher Education. HPLC analysis was carried out with the equipment purchased, thanks to the financial support of the European Regional Development Fund in the framework of the Polish Innovation Economy Operational Program POIG.02.01.00-12-167/08. A method of diadinoxanthin and diatoxanthin purification has been reported as patent applications P.412178 and P.412177 (Kuczynska & Jemiola-Rzeminska, 2015) and PCT/PL2016/000047 (Kuczynska & Jemiola-Rzeminska, 2016).

Author details

Paulina Kuczynska[1], Malgorzata Jemiola-Rzeminska[1,2] and Kazimierz Strzalka[1,2*]

Address all Correspondence to : kazimierzstrzalka@gmail.com

1 Department of Plant Physiology and Biochemistry, Faculty of Biochemistry, Biophysics and Biotechnology, Jagiellonian University, Krakow, Poland

2 Malopolska Centre of Biotechnology, Jagiellonian University, Krakow, Poland

References

[1] Britton G, Liaaen-Jensen S, Pfander H, editors. Carotenoids handbook. 1st ed, Birkhäuser Verlag, Basel; 2004

[2] Vílchez C, Forján E, Cuaresma M, Bédmar F, Garbayo I, Vega JM. Marine carotenoids: Biological functions and commercial applications. Mar Drugs 2011;**9**:319–33. DOI: 10.3390/md9030319

[3] Latowski D, Grzyb J, Strzalka K. The xanthophyll cycle – Molecular mechanism and physiological significance. Acta Physiol Plant 2004;**26**:197–212. DOI: 10.1007/s11738-004-0009-8

[4] García-Plazaola JI, Matsubara S, Osmond CB. The lutein epoxide cycle in higher plants: Its relationships to other xanthophyll cycles and possible functions. Funct Plant Biol 2007;**34**:759–73. DOI: 10.1071/FP07095

[5] Müller P, Li XP, Niyogi KK. Non-photochemical quenching. A response to excess light energy. Plant Physiol 2001;**125**:1558–66. DOI: 10.1104/pp.125.4.1558

[6] Hager A, Holocher K. Localization of the xanthophyll-cycle enzyme violaxanthin de-epoxidase within the thylakoid lumen and abolition of its mobility by a (light-dependent) pH decrease. Planta 1994;**192**:581–9. DOI: 10.1007/BF00203597

[7] Arnoux P, Morosinotto T, Saga G, Bassi R, Pignol D. A structural basis for the pH-dependent xanthophyll cycle in *Arabidopsis thaliana*. Plant Cell 2009;**21**:2036–44. DOI: 10.1105/tpc.109.068007

[8] Saga G, Giorgetti A, Fufezan C, Giacometti GM, Bassi R, Morosinotto T. Mutation analysis of violaxanthin de-epoxidase identifies substrate-binding sites and residues involved in catalysis. J Biol Chem 2010;**285**:23763–70. DOI: 10.1074/jbc.M110.115097

[9] Latowski D, Åkerlund HE, Strzałka K. Violaxanthin de-epoxidase, the xanthophyll cycle enzyme, requires lipid inverted hexagonal structures for its activity. Biochemistry 2004;**43**:4417–20. DOI: 10.1021/bi049652g

[10] Jahns P, Latowski D, Strzalka K. Mechanism and regulation of the violaxanthin cycle: The role of antenna proteins and membrane lipids. Biochim Biophys Acta - Bioenerg 2009;**1787**:3–14. DOI: 10.1016/j.bbabio.2008.09.013

[11] Siefermann D, Yamamoto HY. NADPH and oxygen-dependent epoxidation of zeaxanthin in isolated chloroplasts. Biochem Biophys Res Commun 1975;**62**:456–61. DOI: 10.1016/S0006-291X(75)80160-4

[12] Gilmore AM, Mohanty N, Yamamoto HY. Epoxidation of zeaxanthin and antheraxanthin reverses non- photochemical quenching of photosystem II chlorophyll a fluorescence in the presence of trans- thylakoid delta pH. FEBS Lett 1994;**350**:271–4

[13] Büch K, Stransky H, Hager A. FAD is a further essential cofactor of the NAD (P) H and O 2-dependent zeaxanthin-epoxidase. FEBS Lett 1995;**376**:45–58

[14] Goss R, Böhme K, Wilhelm C. The xanthophyll cycle of *Mantoniella squamata* converts violaxanthin into antheraxanthin but not to zeaxanthin: Consequences for the mechanism of enhanced non-photochemical energy dissipation. Planta 1998;**205**:613–21. DOI: 10.1007/s004250050364

[15] Rmikil N-E, Brunet C, Cabioch J, Lemoine Y. Xanthophyll-cycle and photosynthetic adaptation to environment in macro- and microalgae. Hydrobiologia 1996;**326–327**:407–13. DOI: 10.1007/BF00047839

[16] Stransky H, Hager A. The carotenoid pattern and the occurrence of the light-induced xanthophyll cycle in various classes of algae. VI. Chemo systematic study. Arch Mikrobiol 1970;**73**:315–23

[17] Lohr M, Wilhelm C. Algae displaying the diadinoxanthin cycle also possess the violaxanthin cycle. Proc Natl Acad Sci U S A 1999;**96**:8784–9. DOI: 10.1073/pnas.96.15.8784

[18] Jakob T, Goss R, Wilhelm C. Unusual pH-dependence of diadinoxanthin de-epoxidase activation causes chlororespiratory induced accumulation of diatoxanthin in the diatom *Phaeodactylum tricornutum*. J Plant Physiol 2001;**158**:383–90

[19] Grouneva I, Jakob T, Wilhelm C, Goss R. Influence of ascorbate and pH on the activity of the diatom xanthophyll cycle-enzyme diadinoxanthin de-epoxidase. Physiol Plant 2006;**126**:205–11. DOI: 10.1111/j.1399-3054.2006.00613.x

[20] Goss R, Ann Pinto E, Wilhelm C, Richter M. The importance of a highly active and pH-regulated diatoxanthin epoxidase for the regulation of the PS II antenna function in diadinoxanthin cycle containing algae. J Plant Physiol 2006;**163**:1008–21. DOI: 10.1016/j.jplph.2005.09.008

[21] Mewes H, Richter M. Supplementary ultraviolet-B radiation induces a rapid reversal of the diadinoxanthin cycle in the strong light-exposed diatom *Phaeodactylum tricornutum*. Plant Physiol 2002;**130**:1527–35. DOI: 10.1104/pp.006775

[22] Jahns P, Holzwarth AR. The role of the xanthophyll cycle and of lutein in photoprotection of photosystem II. Biochim Biophys Acta 2012;**1817**:182–93

[23] Raniello R, Lorenti M, Brunet C, Buia MC. Photoacclimation of the invasive alga *Caulerpa racemosa* var. cylindracea to depth and daylight patterns and a putative new role for siphonaxanthin. Mar Ecol 2006;**27**:20–30. DOI: 10.1111/j.1439-0485.2006.00080.x

[24] Latowski D, Kruk J, Burda K, Skrzynecka-Jaskier M, Kostecka-Gugala A, Strzalka K. Kinetics of violaxanthin de-epoxidation by violaxanthin de-epoxidase, a xanthophyll cycle enzyme, is regulated by membrane fluidity in model lipid bilayers. Eur J Biochem 2002;**269**:4656–65. DOI: 10.1046/j.1432-1033.2002.03166.x

[25] Kuczyńska P, Latowski D, Niczyporuk S, Olchawa-Pajor M, Jahns P, Gruszecki WI, et al. Zeaxanthin epoxidation – An *in vitro* approach. Acta Biochim Pol 2012;**59**:105–7

[26] Bouvier F, D'Harlingue A, Hugueney P, Marin E, Marion-Poll A, Camara B. Xanthophyll biosynthesis. J Biol Chem 1996;**271**:28861–7. DOI: 10.1074/jbc.271.46.28861

[27] Pangestuti R, Kim SK. Biological activities and health benefit effects of natural pigments derived from marine algae. J Funct Foods 2011;**3**:255–66. DOI: 10.1016/j.jff.2011.07.001

[28] Dufossé L, Galaup P, Yaron A, Arad SM, Blanc P, Murthy KNC, et al. Microorganisms and microalgae as sources of pigments for food use: A scientific oddity or an industrial reality?. Trends Food Sci Technol 2005;**16**:389–406. DOI: 10.1016/j.tifs.2005.02.006

[29] Ausich RL. Commercial opportunities for carotenoid production by biotechnology. Pure Appl Chem 1997;**69**:2169–74. DOI: 10.1351/pac199769102169

[30] Kuczynska P, Jemiola-Rzeminska M. Isolation and purification of all-trans diadinoxanthin and all-trans diatoxanthin from diatom *Phaeodactylum tricornutum*. J Appl Phycol. DOI: 10.1007/s10811-016-0961-x

[31] Kuczynska P, Jemiola-Rzeminska M, Strzalka K. Photosynthetic pigments in diatoms. Mar Drugs 2015;**13**:5847–81. DOI: 10.3390/md13095847

[32] Guillard RRL. Culture of phytoplankton for feeding marine invertebrates. Cult Mar Invertebr Anim 1975;**1**:29–60. DOI: 10.1007/978-1-4615-8714-9

[33] Hagerthey SE, William Louda J, Mongkronsri P. Evaluation of pigment extraction methods and a recommended protocol for periphyton chlorophyll a determination and chemotaxonomic assessment. J Phycol 2006;**42**:1125–36. DOI: 10.1111/j.1529-8817.2006.00257.x

[34] Rowan K. Extraction and separation of lipid-soluble pigments. Photosynth Pigment Algae, Cambridge University Press, Cambridge; 1989, p. 350

[35] Khachik F, Beecher GR, Whittaker NF. Separation, identification, and quantification of the major carotenoid and chlorophyll constituents in extracts of several green vegetables by liquid chromatography. J Agric Food Chem 1986;**34**:603–16. DOI: 10.1021/jf00070a006

[36] Kaczor A, Baranska M. Extraction and prepreparation of carotenoids. Carotenoids Nutr Anal Technol, John Wiley & Sons, Ltd; 2016, p. 320. DOI: 10.1002/9781118622223.ch1

[37] Huang L, Haug A. Regulation of membrane lipid fluidity in Acholeplasma laidlawii: Effect of carotenoid pigment content. Biochim Biophys Acta - Biomembr 1974;**352**:361–70. DOI: 10.1016/0005-2736(74)90228-4

[38] Gruszecki WI. Carotenoids in membranes. In: Frank H, Young AJ, Britton G, Codgell RJ, editors. Photochem carotenoids, Kluwer Academic Publ, Dordrecht; 1999:pp. 363–79

[39] Milanowska J, Polit A, Wasylewski Z, Gruszecki WI. Interactions of isomeric forms of xanthophyll pigment zeaxanthin with dipalmitoylphosphatidylcholine studied in monomolecular layers. J Photochem Photobiol B Biol 2003;**72**:1–9. DOI: 10.1016/j.jphotobiol.2003.08.009

[40] Gruszecki WI, Sujak A, Strzalka K, Radunz A, Schmid GH. Organisation of xanthophyll-lipid membranes studied by means of specific pigment antisera, spectrophotometry and monomolecular layer technique lutein versus zeaxanthin. Z Fur Naturforsch - Sect C J Biosci 1999;**54**:517–25

[41] Sujak A, Mazurek P, Gruszecki WI. Xanthophyll pigments lutein and zeaxanthin in lipid multibilayers formed with dimyristoylphosphatidylcholine. J Photochem Photobiol B Biol 2002;**68**:39–44. DOI: 10.1016/S1011-1344(02)00330-5

[42] Gruszecki WI, Strzalka K. Carotenoids as modulators of lipid membrane physical properties. Biochim Biophys Acta - Mol Basis Dis 2005;**1740**:108–15. DOI: 10.1016/j.bbadis.2004.11.015

[43] Subczynski WK, Markowska E, Gruszecki WI, Sielewiesiuk J. Effects of polar carotenoids on dimyristoylphosphatidylcholine membranes: A spin-label study. BBA - Biomembr 1992;**1105**:97–108. DOI: 10.1016/0005-2736(92)90167-K

[44] Subczynski WK, Markowska E, Sielewiesiuk J. Spin-label studies on phosphatidylcholine-polar carotenoid membranes: Effects of alkyl-chain length and unsaturation. BBA - Biomembr 1993;**1150**:173–81. DOI: 10.1016/0005-2736(93)90087-G

[45] Subczynski WK, Markowska E, Sielewiesiuk J. Effect of polar carotenoids on the oxygen diffusion-concentration product in lipid bilayers. An EPR spin label study. BBA - Biomembr 1991;**1068**:68–72. DOI: 10.1016/0005-2736(91)90061-C

[46] Kostecka-Gugała A, Latowski D, Strzałka K. Thermotropic phase behaviour of α-dip almitoylphosphatidylcholine multibilayers is influenced to various extents by carotenoids containing different structural features – Evidence from differential scanning calorimetry. Biochim Biophys Acta - Biomembr 2003;**1609**:193–202. DOI: 10.1016/S0005-2736(02)00688-0

[47] Sujak A, Gabrielska J, Grudziński W, Borc R, Mazurek P, Gruszecki WI. Lutein and zeaxanthin as protectors of lipid membranes against oxidative damage: The structural aspects. Arch Biochem Biophys 1999;**371**:301–7. DOI: 10.1006/abbi.1999.1437

[48] Gabrielska J, Gruszecki WI. Zeaxanthin (dihydroxy-beta-carotene but not beta-carotene rigidifies lipid membranes: A H-1-NMR study of carotenoid-egg phosphatidylcholine liposomes. Biochim Biophys Acta - Biomembr 1996;**1285**:167–74. DOI: 10.1016/s0005-2736(96)00152-6

[49] Gruszecki WI, Siewielesiuk J. Orientation of xanthophylls in phosphatidylcholine multibilayers. Biochim Biophys Acta - Biomembr 1990;**1023**:405–412

[50] Augustynska D, Jemiola-Rzeminska M, Burda K, Strzalka K. Influence of polar and nonpolar carotenoids on structural and adhesive properties of model membranes. Chem Biol Interact 2015;**239**:19–25. DOI: 10.1016/j.cbi.2015.06.021

Chapter 2

Electronic Structure of Carotenoids in Natural and Artificial Photosynthesis

Manuel Flores-Hidalgo, Francisco Torres-Rivas,
Jesus Monzon-Bensojo, Miguel Escobedo-Bretado,
Daniel Glossman-Mitnik and Diana Barraza-Jimenez

Additional information is available at the end of the chapter

http://dx.doi.org/10.5772/67636

Abstract

This chapter is about a theoretical study applied to six carotenoids present in vegetables containing carotenes and xanthophylls. Electronic properties are analyzed such as energy in frontier orbitals and the first molecular orbitals to work in the UV-Vis absorption spectroscopy. Electronic structure methodologies were used within the frame of the density functional theory (DFT) using the theoretical methods B3LYP/6-31G(d)//B3LYP/6-31G+(d,p) for ground states and B3LYP/6-31G(d)//CAM-B3LYP/6-31G+(d,p) for excited states. Results for the main absorption peak are in agreement with experimental results with a difference between zeaxanthin and violaxanthin results of 0.1 eV, approximately. The UV-Vis absorption spectra obtained for carotenoids are in good agreement with the experimental results. The possible use in energy generation systems is discussed for these systems. Diade chlorophyllide *a*-zeaxanthin was formed, and calculation results predicted energy transfer for these photosynthetic systems.

Keywords: DFT, artificial photosynthesis, carotenoids, xanthophylls, diade chl *a*-zx

1. Introduction

Natural photosynthesis requires the participation of chlorophyll *a* and accessory pigments. Carotenoids are the more commonly used accessory pigments. In photosynthesis, plants and organisms convert light energy into chemical energy that can later be released to fuel organisms' activities; therefore, it is an energy transformation. It is one of the principal processes in nature and it is fundamental for life existence. Solar energy conversion to chemical fuels using green methodologies may be approached with photosynthesis [1] since this natural process is the main user of solar energy in our planet. This natural process uses effectively the largest exploitable renewable energy resource. Solar energy provides our planet with more energy per hour than the total energy consumed by human activities in 1 year. In other words, direct conversion of solar energy into chemical fuels represents an optimal approach to address the globally growing energy demand in a sustainable way [1–2]. Photosynthesis if reproduced may address a lot of our environmental problems derived from energy conversion.

In this way, mimicking photosynthesis has become a subject of great interest in the scientific world, and this global research trend has given origin to a recently created term, artificial photosynthesis [1–3]. This concept refers to a chemical process that replicates the natural process of photosynthesis; it mainly studies the process to convert sunlight, water, and carbon dioxide into carbohydrates and oxygen. This process aims to emulate natural ways by using man-made devices to convert and store solar energy using chemical fuels as feedstock [3]. To absorb the visible light part of the solar radiation (350–700 nm), green plants use chlorophyll *a* as the main light absorber along with a number of accessory pigments such as xanthophylls, carotenoids, and a modified form of chlorophyll, called chlorophyll *b*. Chlorophyll *a* absorbs in the blue-violet, orange-red spectral regions while the accessory pigments cover the intermediate yellow-green-orange part [3–4].

Carotenoids are important in photosynthesis, and with the mimicking of this natural process, they have raised their importance due to the fundamental need for renewable energy sources such as artificial photosynthesis [5]. There are other fields in which carotenoids are important as well, such as food or health. Fruits and vegetables are the principal sources of carotenoids and play an important role in diet due to vitamin A activity [5–6]. In addition to this, carotenoids are also important for antioxidant activity, intercellular communication, and immune system activity [6–8]. Epidemiological studies reported that the consumption of diets rich in carotenoids is associated with a lower incidence of cancer, cardiovascular diseases, age-related macular degeneration, and cataract formation [9–10]. Deficiency of carotenoids results in clinical signs of conjunctiva and corneal aberrations, including xerophthalmia, night blindness, corneal ulceration, scarring, and resultant irreversible blindness [11].

Carotenoids are classified in carotenes and xanthophylls. Carotenes contain only a parent hydrocarbon chain without any functional group, such as α-carotene, β-carotene, and lycopene. Xanthophylls contain oxygen as the functional group, including lutein and zeaxanthin [12]. In plant tissues, carotenoids are typically located in chromoplasts (specific organelles) inside cells. Substructures composed of lipids, proteins, and carotenoids are being synthesized during chromoplast development, and depending on their morphology, they can be classified as crystalline, globular, fibrillar, membranous, or tubular-type chromoplasts [13–15]. Carotenoids are embedded in a complex structural organization. Carotenes and xanthophylls contain more than seven conjugate bonds that enable visible light absorption and from here, they have the capability to participate in the photosynthesis [6]. For light energy to be transformed into chemical energy, the electronic structure is fundamental in understanding how natural photosynthesis occurs and how this process can be associated with clean energy generation through artificial photosynthesis. Due to their chemical structure, carotenoids are tetraterpenoids so they have a long chain of conjugated double bonds; for this reason, these micronutrients are highly lipophilic [13, 16–18].

There are different determination methods to find out the basic chemical structure of carotenoids. Their structure is based on eight isoprenoid units with a conjugated double-bond system, which makes isomeric forms very common [19]. In addition, double bonds in the carbon chain make carotenoids susceptible to reactions, such as oxidation and isomerization (cis-trans), especially due to light, heat, acids, and oxygen [20]. Cyclization, hydrogenation, dehydrogenation, or additions of lateral groups, among others, are some modifications that lead to an extremely complex variety of compounds with common structures [19].

Moreover, carotenoids can be found in nature both in their free form and also in a more stable, esterified form with fatty acids [21]. The high variability in their chemical structure and their poor stability greatly contribute to the difficulty of carotenoid analysis. Also, there is a lack of commercially available standards and other important reasons [22] that make it difficult to have more analytical methods to identify and to measure carotenoids in real samples [21–23].

Another analytic alternative is related to theoretical methods. There is a wide chart of choices to model and simulate these compounds that range from macro-, micro-, and atomic to subatomic methodologies. In this work, we use density functional theory (DFT) to learn more about carotenoids. We use DFT to model and simulate carotenoids' ground states as well as excited states and analyze their electronic structure. Fundamentally, carotenoids have a strong absorption of visible light in the blue and green region of the solar spectrum. Most carotenoids found in photosynthetic organisms have the characteristic colors yellow, orange, and red. The lowest excited single state in most pigment molecules represents the lowest energy, which optically allows a one-photon transition from the ground state [13]. This chapter provides numerical data to parameterize some of the more important properties of carotenoids. It provides a good insight about their important role in both natural and artificial photosynthesis, and since these results relate to its more basic features, it can be useful for other applications as well, such as in the food and health industries.

2. Computational methods and details

All calculations were carried out employing Gaussian 09 program suite [24]. This chapter was developed with computational calculations employing electronic structure methods using density functional theory (DFT). Then, a vibrational frequencies' calculation was carried out to corroborate a global minimum. These calculations, geometry optimization, and vibrational frequencies were performed in the gaseous phase, using a methanol-like solvent. Molecular orbitals for the different carotenoids were obtained with energy calculations using the B3LYP/6-31G(d)//B3LYP/6-31+G(d,p) theoretical method [25]. Excited states in the gaseous phase were carried out for all six carotenoid variants within this work and in the solvent phase for xanthophylls molecules. These later calculations allowed us to obtain molecular orbitals and absorption states. The same set of calculations used in carotenoids was applied to chlorophyll *a* in the gaseous phase, with the objective of forming one of the main diades that has been found as a participant in natural photosynthesis. CAM-B3LYP [26] functional was used in all excited states' calculations using the time-dependent density functional theory (TDDFT). Molecular orbitals data was processed to obtain orbitals' diagrams and the absorption spectra with Chemissan code [27].

3. Results and discussions

In this section, calculations results are displayed and analyzed. The first set of results contains ground states data; first, the geometric structures are displayed and next the energy results are displayed, including molecular orbitals, energy gap, and relevant chemical properties. In the second set of results are included excited states data with their corresponding molecular orbital diagrams and absorption spectra based on TDDFT calculations.

3.1. Carotenoid structures

Carotenoids included in this work are displayed in **Figure 1**. Beta-carotene and lycopene are carotenes with the characteristic of belonging to the hydrocarbons group, which means that their structure includes only carbon and hydrogen atoms. The rest are four carotenoids that belong to the xanthophylls which characterize themselves by containing within their structure carbon and hydrogen with oxygen atoms bonded to the six-carbon ring.

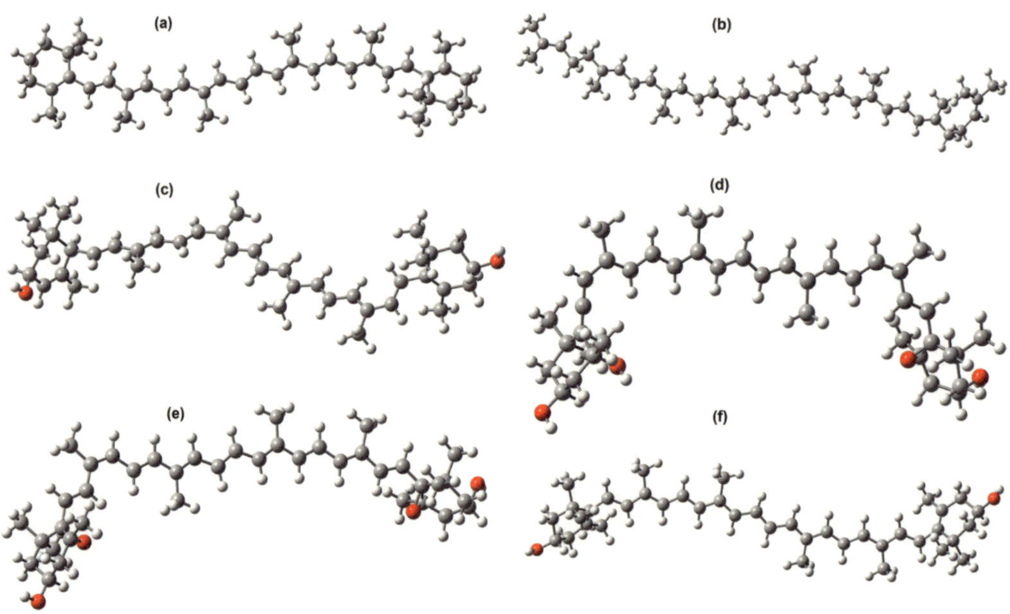

Figure 1. Geometry optimization of carotenoid structures: (a) Beta-carotene, (b) lycopene, (c) lutein, (d) neoxanthin, (e) violaxanthin, (f) zeaxanthin.

http://dx.doi.org/10.5772/67636

3.1.1. Carotenoids ground states

The ground states energy results allow us to obtain data related to electronic conduction capabilities of the selected molecules in their ground states. For these calculations, the interpretation scheme of the difference between the Highest Occupied Molecular Orbital (HOMO) and the Lowest Unoccupied Molecular Orbital (LUMO) was applied, where HOMO and LUMO are frontier orbitals located in the valence band and in the conduction band, respectively. Resulting values for these orbitals relate to ionization energy in the case of HOMO, which means that a lower ionization energy corresponds to a higher HOMO energy. In fact, a lower energy for LUMO is associated with a higher electronic affinity. These effects were explained in more detail in our published work [28] and can be studied by further reading **Table 1** as shown ahead in this section.

Frontier molecular orbitals' values are shown in **Table 1**, and **Figure 2** displays molecular orbitals obtained for the six selected carotenoid structures in their gaseous phase that will be discussed next. The diagram is divided into two sections: the first belongs to carotenes and the second to xanthophylls. The HOMO-LUMO energy gap is obtained by calculating the difference between frontier orbitals' energy values. Beta-carotene and lycopene energy gap results have a small difference. Then, it is observed that these compounds present a similar trend in their HOMO orbitals. Based on these results, xanthophylls have a higher ionization energy which derives in the capability present in one of these variants to form a diade integrated system configured by chlorophyll *a*-xanthophyll.

Carotenoids	Gas/solvent	IP	EA	HOMO	LUMO	B3LYP (HOMO-LUMO)*
β-Carotene	Gas	5.41	1.549	-4.529	-2.443	2.086
	Methanol	4.504	2.709	-4.636	-2.558	2.078
Lycopene	Gas	5.429	1.597	-4.553	-2.487	2.066
	Methanol	4.533	2.755	-4.668	-2.605	2.063
Lutein	Gas	5.665	1.502	-4.699	-2.458	2.241
	Methanol	4.615	2.678	-4.753	-2.524	2.229
Neoxanthin	Gas	5.652	1.37	-4.672	-2.347	2.325
	Methanol	4.668	2.633	-4.805	-2.483	2.322
Violaxanthin	Gas	5.744	1.484	-4.772	-2.459	2.313
	Methanol	4.716	2.684	-4.852	-2.539	2.313
Zeaxanthin	Gas	5.594	1.555	-4.68	-2.487	2.193
	Methanol	4.61	2.71	-4.744	-2.562	2.182

Results are in electron Volts (eV).
*Absolute values.

Table 1. Global chemical reactivity parameters (IP and EA), energy levels (HOMO/LUMO), and energy gap (HOMO-LUMO) in the gaseous phase, and methanol as a solvent for all B3LYP/6-31+G(d,p)-analyzed carotenoids.

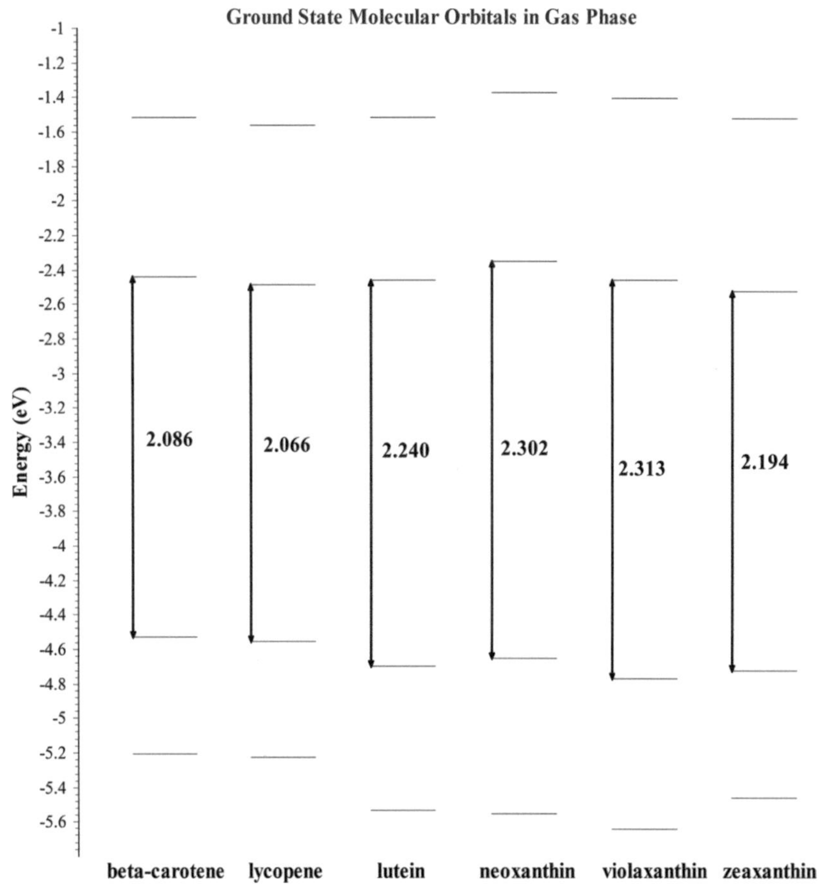

Figure 2. Ground states molecular orbitals for selected carotenoid structures. The calculation was carried out in the gaseous phase using DFT with the B3LYP/6-31+G(d,p) theoretical method.

From xanthophylls displayed in the molecular orbitals diagram, one can see that zeaxanthin has the narrower HOMO-LUMO energy gap. A narrow HOMO-LUMO energy difference benefits energy transfer process.

Photosynthesis requires that plants containing chlorophyll *a* capture light to transform it into chemical energy. The role that light plays consists of producing a luminous excitation that impacts an electron, allowing this charged particle to jump from a ground state of inferior energy to an excited state with higher energy and later return to the lower energy state. This process is known as excited states in the electronic structure.

In photosynthesis, carotenoids play the role of the aforementioned accessory pigments. Excited states analysis explains xanthophylls' performance as pigments that are part of the photosynthetic process.

Figure 3 displays a molecular orbital diagram for excited states corresponding to all molecules within our chapter. This diagram, as occurred for ground states, provides information discussed previously in this section.

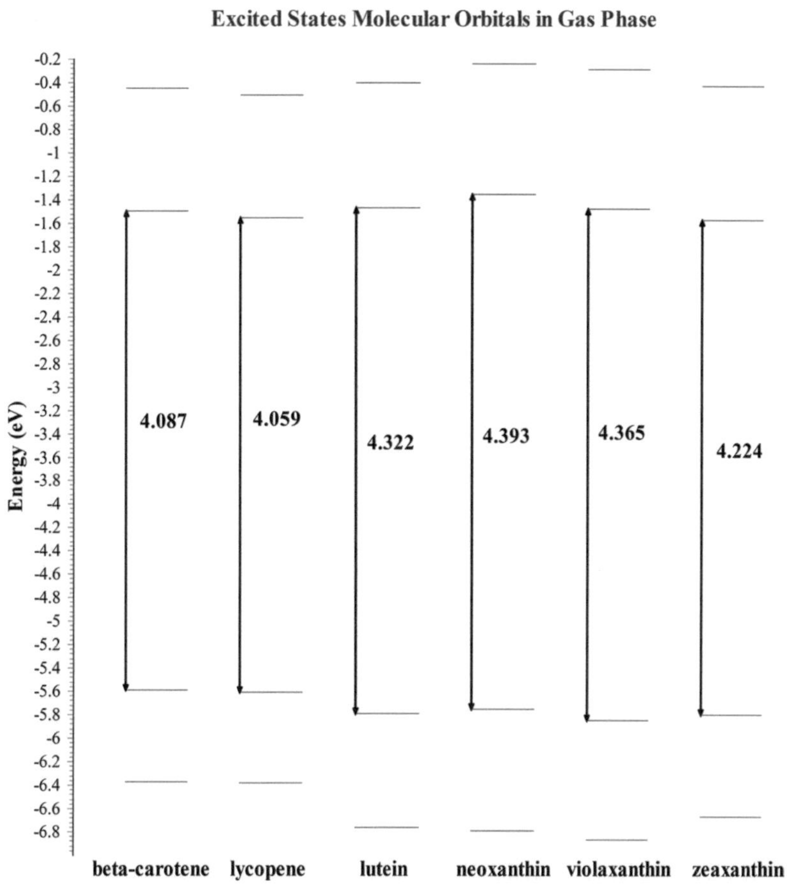

Figure 3. Excited states molecular orbitals for selected carotenoid structures. The calculation was carried out in the gaseous phase using TDDFT with the CAM-B3LYP/6-31+G(d,p) theoretical method.

3.1.2. *Carotenoids excited states molecular orbitals*

As occurs in ground states, in excited states, zeaxanthin is the molecule with the narrower energy gap from the xanthophylls-type carotenoids which means this is the molecule that can be excited more easily. Other data important to consider in excited states analysis is the UV-Vis absorption spectra. Absorption spectra enable us to identify the wavelength in which a pigment absorbs sunlight and thus locate the reference electromagnetic radiation working range and the visible light required to favor photosynthesis.

Now, depending on the solvent used, sunlight absorption may have a benefit. In general, the absorption extends in larger wavelength when the solvent is employed if compared to the absorption results in the gaseous phase. **Figure 4** displays the diagram of excited states molecular orbitals with the use of solvents, which in this case is methanol. This excited states calculation was carried out only for xanthophylls because zeaxanthin is the accessory pigment used to form the diade with chlorophyll *a*, considering that the latter is the main photosynthetic pigment. Molecular orbitals shown so far, both in the gaseous phase and in the solvent phase correspond to those involved in the main absorption peak found.

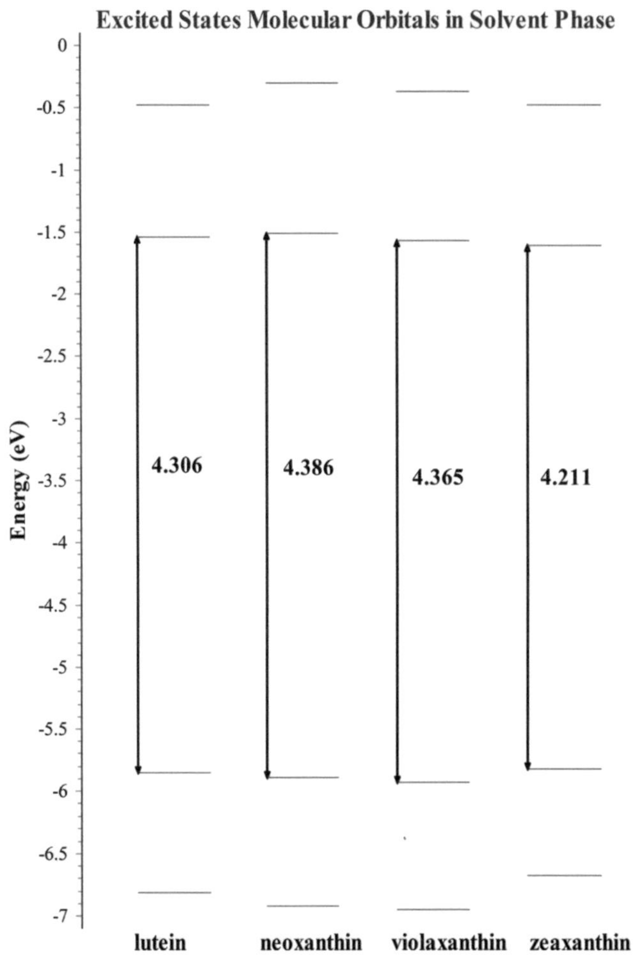

Figure 4. Excited states molecular orbitals for selected carotenoid structures. Calculations were carried out in methanol as a solvent with TDDFT using the CAM-B3LYP/6-31+G(d,p) theoretical method.

If one observes the molecular orbitals diagram corresponding to the molecules in methanol as a solvent, one can see that zeaxanthin is the molecule with the narrower energy gap as occurred in the gaseous phase calculations. Another important observation is how violaxanthin keeps the same values in both calculations, in the gaseous phase and in the solvent phase.

3.1.3. Carotenoids excited states UV-Vis spectra

Absorption spectra for carotenoids-type beta-carotene and lycopene for the gaseous phase are shown in **Figure 5**. In this figure, a high coincidence in the absorption maximum peak between beta-carotene with 472 nm and lycopene with 476 nm is observed. According to these calculations, the maximum absorption peak for beta-carotene is surpassed by 20 nm. Meanwhile for lycopene, a better approximation with the experimental value of 470 nm is obtained.

http://dx.doi.org/10.5772/67636

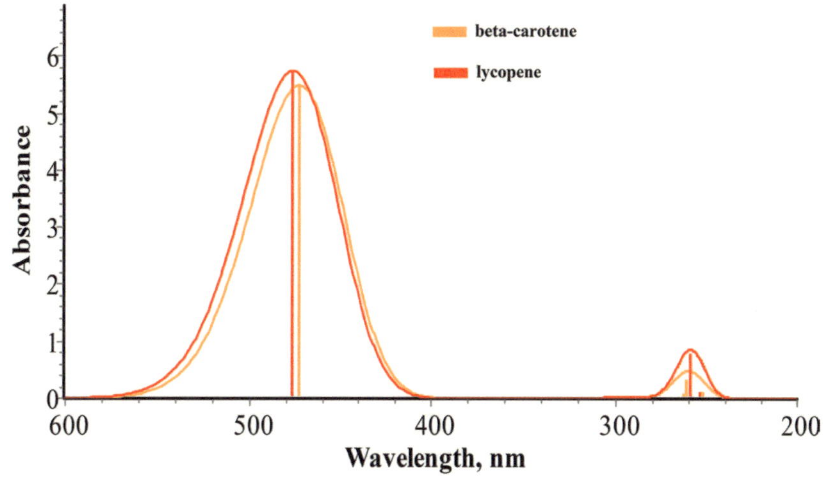

Figure 5. UV-Vis absorption spectra for carotenoids-type beta-carotene and lycopene. Calculations were carried out with TDDFT using the CAM-B3LYP/6-31+G(d,p) theoretical method.

3.1.4. Xanthophylls excited states UV-Vis spectra

Absorption spectra for the gaseous phase and with the solvent for xanthophylls-type carotenoids are displayed in **Figure 6**. According to the figures, one can observe that the variation in absorption results in the gaseous phase with respect to the solvent phase is different only by a wavelength displacement. Zeaxanthin is the carotenoid from xanthophylls with an absorption in a longer wavelength, similar to lutein. The difference between them is the absorbance value where zeaxanthin has the bigger absorbance.

3.2. Diades

Diades structures are formed with two molecules, one from the group of the main photosynthetic pigments and the other from the accessory pigments group. For this work, we used chlorophyll *a*-carotenoid. In the formation of the system, chlorophyll *a*-zeaxanthin was employed, and the chlorophyllide *a,* due to the phytol, lacks contribution in the photochemical activity related to light absorbance in chlorophyll *a* but brings about some benefit with some computational cost reduction. In the next paragraphs, we discuss our results for the selected diade. Our discussion for these systems is a relative view between ground states and excited states that will enable us to understand their electronic structure properties. Our discussion for this system is a relative view between ground states and excited states that will enable us to understand their electronic structure properties.

For the intermolecular energy transfer study, photosynthetic pigments' diade formed by chlorophyll *a*-zeaxanthin was modeled and analyzed. For construction of the diade system, we used chlorophyllide *a,* and zeaxanthin was centered, and to build this diade, the individual optimized structures were used. Once the diade was modeled, the system was subject to a geometric optimization using the B3LYP/6-31G (d) theoretical method in the gaseous phase.

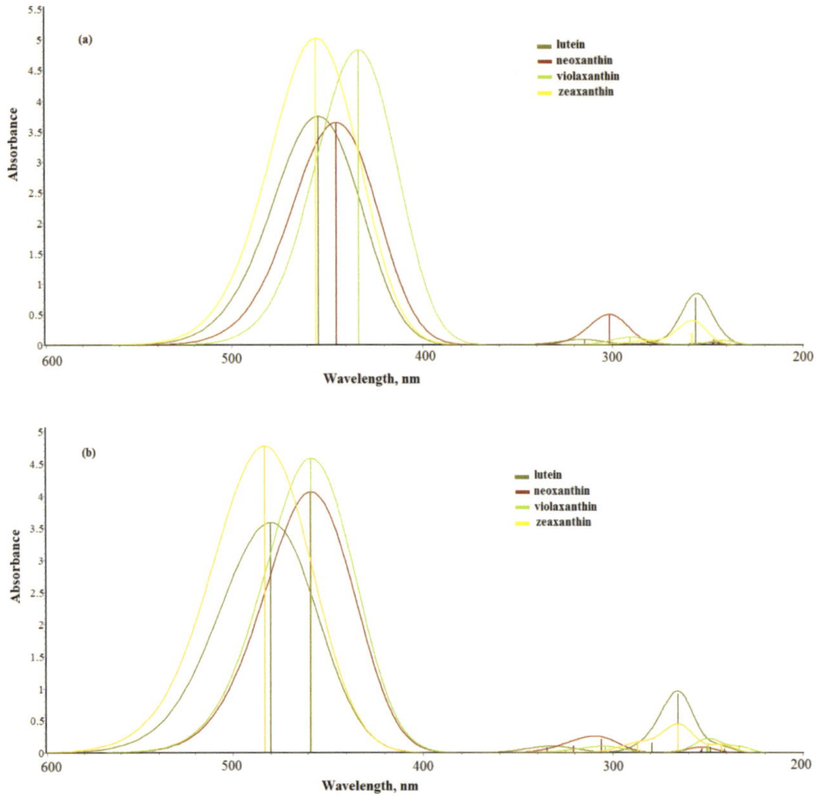

Figure 6. UV-Vis absorption spectra for xanthophylls-type carotenoid structures. Calculations are in (a) the gaseous phase and (b) by using methanol as a solvent with TDDFT using the CAM-B3LYP/6-31+G(d,p) theoretical method.

For the determination of the intermolecular distances between chlorophyllide *a* and zeaxanthin, we performed a series of prior work to determine the best choice with distances between 5.0 and 9 Å (all calculations included vibrations frequencies analysis to make sure that the global minima was reached).

Since electronic properties of selected photosynthetic pigments depend of nature in both cases, the ground and the excited states, we moved forward with molecular orbitals analysis for isolated molecules and diade systems (in both ground and excited states). TDDFT was used, with Tamm-Dancoff approximation (TDA) and the CAMB3LYP/6-31+G(d,p) theoretical method.

After HOMO/LUMO energy analysis, UV-Vis absorption spectra were obtained for isolated molecules and their corresponding diade systems. The same theoretical level was used with same functional set.

3.2.1. Diade molecular orbitals (HOMO/LUMO energies)

Results for molecular orbitals and HOMO-LUMO energies corresponding to chlorophyll *a*, zeaxanthin, and the respective diade formed by them, in their ground and excited states, are shown in **Figure 7(a)** and **(b)**, respectively. This way of organizing the diagrams enables one

http://dx.doi.org/10.5772/67636

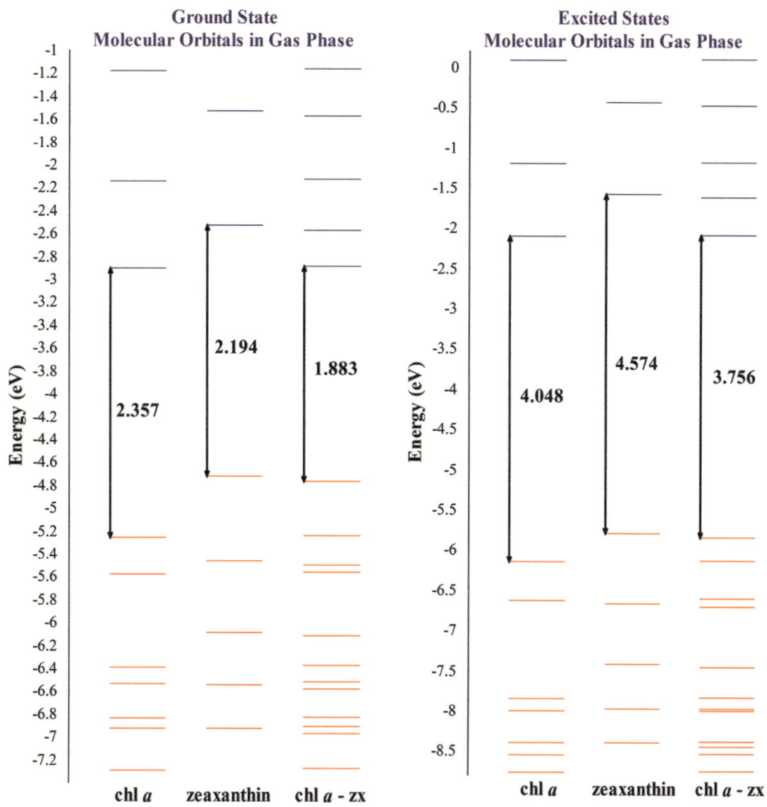

Figure 7. Molecular orbitals for chlorophyllide *a*, zeaxanthin, and their corresponding diade in both ground and excited states. Calculations for ground states were obtained using DFT with B3LYP/6-31+G(d,p) and excited states were obtained using the TDDFT scheme with TDA approximation employing the CAMB3LYP/6-31+G(d,p) theoretical method. Energies in eV.

to compare both, ground and excited states, and obtain an easy way to compare the results. For the specific case of diade formed by chlorophyllide *a* and zeaxanthin in its ground state, when analyzed separately, chlorophyllide *a* presented an energy gap of 2.357 eV. When the diade was analyzed, the energy gap decreased to 1.883 eV. Meanwhile, at the excited state, chlorophyllide *a's* result had 4.048 eV (isolated), and for the diade, it was obtained as 3.756 eV.

Figure 7 shows an energy gap for chlorophyllide *a* which decreased in both energetic states (ground and excited), presenting a difference between the isolated chlorophyllide *a* and those integrated into a diade system of 0.474 and 0.292 eV, respectively. This indicates in a general way that chlorophyllide *a* has a higher reactivity when it is found in the diade system, and this is independent of the energetic state.

For the diade system, it was found that the HOMO molecular orbital is provided by the corresponding carotenoid zeaxanthin, while the LUMO orbital corresponds to chlorophyllide *a*. **Figure 8** displays both HOMO and LUMO orbitals for diade chlorophyllide *a*/zeaxantina, indicating in general that carotenoids are responsible for luminous energy absorption, by means of HOMO molecular orbitals, and interact with LUMO from chlorophyllide *a*.

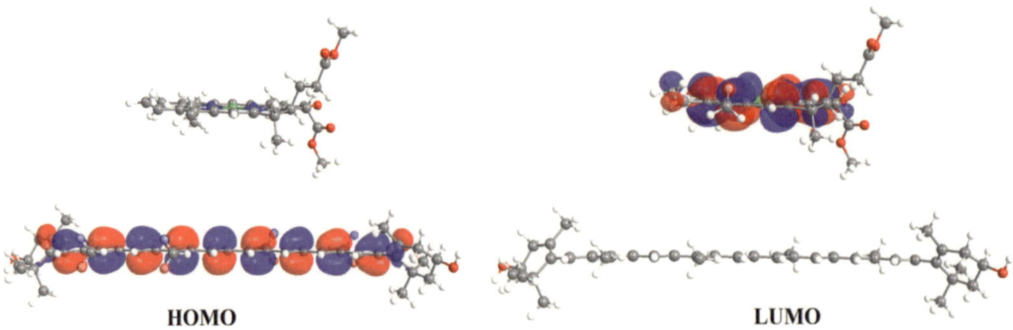

Figure 8. Molecular orbitals HOMO and LUMO's graphical representations for the diade systemchlorophyllide *a*/zeaxanthin.

3.2.2. Absorption UV-vis spectra for chlorophyllide a and chlorophyll a

From excited states energy calculations, UV-vis absorption spectra were determined for each of the two selected molecules including integrated systems in the diade. In **Figure 9** are displayed UV-vis absorption spectra for experimental and theoretical (corresponding to experimentation similar to calculations within this work) frameworks for chlorophyll *a*. Chlorophyllide *a* presents a similar structure than chlorophyll *a*, except that in chlorophyllide *a*, the phytol lateral chain is removed which does not present any π bonding (conjugate double bonds). Therefore, it does not present photochemical reactivity [29]. In this way, the

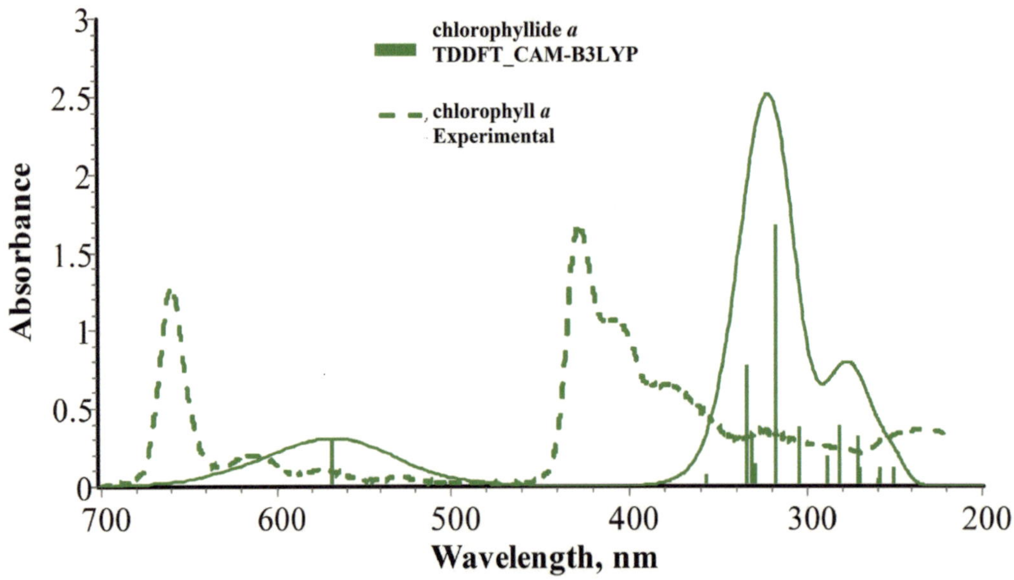

Figure 9. UV-Vis absorption spectra for both chlorophyll *a* (experimental data) and for chlorophyllide *a*. Chlorophyll *a* is analyzed in diethylic ether and chlorophyllide *a* is analyzed in the gaseous phase using TDDFT with CAM-B3LYP/6-31+G(d,p).

comparison between experimental data for chlorophyll *a* (with the phytol lateral chain) and theoretical data for chlorophyllide *a* (without phytol) is in agreement.

In **Figure 9** are shown experimental values for the more characteristic chlorophyll *a* absorption bands [30], which are 670 y 420 nm for Qy and Soret bands, respectively. For chlorophyllide *a*, the theoretical values found were 568 y 320 nm, respectively.

As can be seen in **Figure 9**, chlorophyllide *a* presents characteristic absorption maximum peaks. These peaks are displaced of approximately 100 nm in chlorophyllide *a* with respect to chlorophyll *a* values. This difference between experimental and theoretical values may be attributed mainly to the solvent effect. As mentioned in this section, to obtain experimental values, the samples used diethylic ether as a solvent while for theoretical values, and calculations were made in the gaseous phase. Another contributor is the methodological difference, in concrete, in the computational method employed (which relies on approximations). However, characteristic absorption peaks for these molecules were consistent in all cases.

Regarding theoretical UV-Vis absorption spectra (theoretical), calculations were developed for the two molecules integrated in the diades, and their diagram was based on the excited states' results. In **Figure 10**, chlorophyllide *a* absorption spectra, zeaxanthin, and their corresponding diades are displayed. Here, one can observe that Qy bands for isolated chlorophyllide *a* and the corresponding band for the diade are practically identical (in both axes). Meanwhile, the Soret band corresponding to the diade presents absorption values slightly over the corresponding values for chlorophyllide *a* alone (the values being the same for the wavelength in both cases). Concretely for zeaxanthin in the diade, this pigment contributes to a slight displacement in the x-axis of 468.6 a 470.5 nm with respect to isolated zeaxanthin. Furthermore,

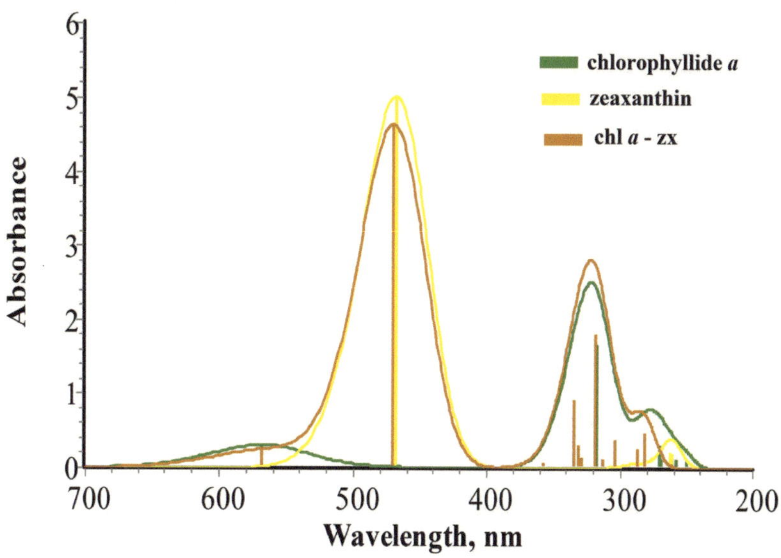

Figure 10. UV-Vis absorption spectra for chlorophyllide *a*, zeaxanthin, and their corresponding diade (chlorophyllide *a*/zeaxanthin). Calculations were carried out under the TDDFT scheme using the CAMB3LYP/6-31+G(d,p) theoretical method.

isolated zeaxanthin presents in this particular peak an absorbance slightly above with respect to the one measured in its corresponding diade. Then, it may be observed how integrated diade systems can cover a wider absorption spectra than isolated chlorophyll *a*.

4. Conclusions

Theoretical calculations allowed us to predict UV-Vis absorption spectra for carotenoid structures selected for this work, achieving a good accuracy with experimental results.

Within the selected carotenoids that were analyzed, zeaxanthin is the pigment with better electronic properties to form an integrated system, chlorophyll *a*-carotenoid.

Excited states molecular orbitals analysis for the diade system formed by chlorophyllide *a*-zeaxanthin allowed one to observe the specific role played by chlorophyll *a* as a photosynthetic pigment by its LUMO contribution to the integrated system on the one hand and on the other hand, the role of zeaxanthin as an accessory pigment with the contribution made to HOMO.

Analysis of pigments related to the natural photosynthetic process allowed us to set a basis on how these pigments can be used in potential artificial photosynthetic applications and to develop new materials for alternative energy applications.

Electronic structure data are fundamental information for any material and may help to develop systems with carotenoids and accessory pigments in other fields such as the health or food industries.

Acknowledgements

The authors thank the National Council of Science and Technology (CONACYT, for its acronym in Spanish) for the financial support in the development of this scientific research to the basic science project, No. 158307. FTR and JMB thank CONACYT for a scholarship to support their doctoral studies.

Author details

Manuel Flores-Hidalgo[1], Francisco Torres-Rivas[2], Jesus Monzon-Bensojo[2], Miguel Escobedo-Bretado[1], Daniel Glossman-Mitnik[3] and Diana Barraza-Jimenez[1,2*]

*Address all correspondence to: dianabarraza@ujed.mx

1 Department of Chemical Sciences, Juarez University of Durango State, Durango, Mexico

2 Food and Development Research Center, A. C. Delicias Unit, Chih, Mexico

3 Advanced Materials Research Center, Chihuahua, Chih, Mexico

References

[1] Su J, Vayssieres L. A place in the sun for artificial photosynthesis? ACS Energy Lett 2016; **1:** 121-135. doi:10.1021/acsenergylett.6b00059

[2] Demmig-Adams B, Stewart JJ, Burch TA, Adams WW, III. Insights from placing photosynthetic light harvesting into context. J Phys Chem Lett 2014; **5**: 2880-2889. doi:10.1021/jz5010768

[3] Tachibana Y, Vayssieres L, Durrant JR. Artificial photosynthesis for solar water-splitting. Nat Photonics 2012; **6:** 511-518. doi:10.1038/nphoton.2012.175

[4] Kalyanasundaram K, Graetzel M. Artificial photosynthesis: biomimetic approaches to solar energy conversion and storage. Curr Opin Biotechnol 2010; **21**: 298–310. doi:10.1016/j.copbio.2010.03.021

[5] Saini RK, Nile SH, Park SW. Carotenoids from fruits and vegetables: chemistry, analysis, occurrence, bioavailability and biological activities. Food Res Int 2015; **76**: 735–750. doi:10.1016/j.foodres.2015.07.047

[6] Haskell MJ. Provitamin A carotenoids as a dietary source of vitamin A. In: Tanumihardjo S.A, editors. Carotenoids and Human Health. 1st ed. New York: Springer; 2013. p. 249–260. doi: 10.1007/978-1-62703-203-2_15

[7] Skibsted LH. Carotenoids in antioxidant networks colorants or radical scavengers. J Agric Food Chem 2012; **60**(10)**:** 2409–2417. doi:10.1021/jf2051416

[8] Stephensen CB. Provitamin A carotenoids and immune function. In: Tanumihardjo S.A, editors. Carotenoids and Human Health. 1st ed. New York: Springer; 2013. p. 261-270. doi: 10.1007/978-1-62703-203-2_16

[9] Meyers KJ, Mares JA, Igo RP, Truitt B, Liu Z, Millen AE, et al. Genetic evidence for role of carotenoids in age-related macular degeneration in the carotenoids in age-related eye disease study (CAREDS). Invest Ophthalmol Vis Sci 2014; **55**(1): 587–599. doi:10.1167/iovs.13-13216

[10] Sharoni Y, Linnewiel-Hermoni K, Khanin M, Salman H, Veprik A, Danilenko M, et al. Carotenoids and apocarotenoids in cellular signaling related to cancer: a review. Mol Nutr Food Res 2012; **56**(2): 259–269. doi:10.1002/mnfr.201100311.

[11] Sommer A. Vitamin A deficiency and clinical disease: a historical overview. J Nutr 2008; **138**(10): 1835–1839.

[12] Namitha KK, Negi PS. Chemistry and biotechnology of carotenoids. Crit Rev Food Sci Nutr 2010; **50**(8): 728–760. doi:10.1080/10408398.2010.499811

[13] Lemmens L, Colle I, Van Buggenhout S, Palmero P, Van Loey A, Hendrickx M. Carotenoid bioaccessibility in fruit- and vegetable-based food products as affected by product (micro) structural characteristics and the presence of lipids: a review. Trends Food Sci Technol 2014; **38**: 125–135. doi:10.1016/j.tifs.2014.05.005

[14] Bartley GE, Scolnik PA. Plant carotenoids: pigments for photoprotection, visual attraction, and human health. Plant Cell 1995; **7**: 1027–1038. doi:10.1105/tpc.7.7.1027

[15] Vishnevetsky M, Ovadis M, Vainstein A. Carotenoid sequestration in plants: the role of carotenoid-associated proteins. Trends Plant Sci 1999; **4**: 232–235. doi:10.1016/S1360-1385(99)01414-4

[16] Britton G. Structure and properties of carotenoids in relation to function. FASEB J 1995; **9:** 1551–1558.

[17] Rodriguez-Amaya DB, Kimura M. Harvest plus handbook for carotenoid analysis. Washington, DC: International Food Policy Research Institute and International Center for Tropical Agriculture; 2004.

[18] Hashimotoa H, Sugaia Y, Uragamia C, Gardiner AT, Cogdell RJ. Natural and artificial light-harvesting systems utilizing the functions of carotenoids. J Photochem Photobiol C Photochem Rev 2015; **25**: 46–70. doi:10.1016/j.jphotochemrev.2015.07.004

[19] Amorim-Carrilho KT, Cepeda A, Fente C, Regal P. Review of methods for analysis of carotenoids. Trends Anal Chem 2014; **56**: 49–73. doi:10.1016/j.trac.2013.12.011

[20] Provesi JG, Dias CO, Amante ER. Changes in carotenoids during processing and storage of pumpkin puree. Food Chem 2011; **128**: 195–202. doi:10.1016/j.foodchem.2011.03.027

[21] Herrero M, Cacciola F, Donato P, Giuffrida D, Dugo G, Dugo P, et al. Serial coupled columns reversed-phase separations in high-performance liquid chromatography: tool for analysis of complex real samples. J Chromatogr A 2008; **1188**: 208–215. doi:10.1016/j.chroma.2008.02.039

[22] Van Breemen RB, Dong L, Pajkovic ND. Atmospheric pressure chemical ionization tandem mass spectrometry of carotenoids. Int J Mass Spectrom 2012; **312**: 163–172. doi:10.1016/j.ijms.2011.07.030.

[23] Cacciola F, Donato P, Beccaria M, Mondello L. Advances in LC-MS for food analysis. LC GC Europe 2012; **25**(5):15–24

[24] Frisch MJ, et al. Gaussian 09, Revision A.1. Wallingford CT: Gaussian Inc.; 2009.

[25] Becke AD. Density functional thermochemistry-III-the role of exact exchange. J Phys Chem A 1993; **98**(7): 5648–5652. doi:10.1063/1.464913

[26] Zhao Y, Truhlar DG. Density functionals with broad applicability in chemistry. Acc Chem Rev 2008; **41**: 157–167. doi:10.1021/ar700111a

[27] Chemissian, A Computer Program to Analyse and Visualise Quantum-Chemical Calculations (L. Skripnikov, 2012).

[28] Torres-Rivas F, Flores-Hidalgo MA, Glossman-Mitnik D, Barraza-Jimenez D. Geometric description and electronic properties of the principal photosynthetic pigments of higher plants: a DFT study. J Mol Model 2015; **21**: 256. doi:10.1007/s00894-015-2796-9

[29] Sundholm D. Comparison of the electronic excitation spectra of chlorophyll *a* and pheophytin a calculated at density functional theory level. Chem Phys Lett 2000; **317**: 545–552. doi:10.1016/S0009-2614(99)01428-1

[30] Strain HH, Thomas MR, Katz JJ. Spectral absorption properties of ordinary and fully deuteriated chlorophylls *a* and *b*. Biochim Biophys Acta 1963; **75**: 306–311. doi:10.1016/0006-3002(63)90617-6

Chapter 3

Localizing and Quantifying Carotenoids in Intact Cells and Tissues

Jerilyn A. Timlin, Aaron M. Collins,
Thomas A. Beechem, Maria Shumskaya and
Eleanore T. Wurtzel

Additional information is available at the end of the chapter

http://dx.doi.org/10.5772/68101

Abstract

Raman spectroscopy provides detailed information about the molecular structure of carotenoids. Advances in detector sensitivity and acquisition speed have driven the expansion of Raman spectroscopy from a bulk analytical tool to a powerful method for mapping carotenoid abundance in cells and tissues. In many applications, the technique is compatible with living organisms, providing highly specific molecular structure information in intact cells and tissues with subcellular spatial resolution. This leads to spatial-temporal-chemical resolution critical to understanding the complex processes in the life cycle of carotenoids and other biomolecules.

Keywords: vibrational spectroscopy, multivariate analysis, confocal Raman microscopy, Raman imaging, photosynthetic organisms, carotenoids, resonance Raman scattering

1. Introduction

Carotenoids are tetraterpenoids and have various functions in plants and algae. They extend the range of wavelengths for photosynthesis by harnessing solar radiation where chlorophyll pigments do not appreciably absorb and serve as structural elements within the photosynthetic apparatus [1]. They also function to dissipate excess solar radiation and prevent the formation of harmful singlet oxygen species [2]. Additionally, in specific photosynthetic eukaryotes, carotenoids accumulate in plastid organelles called chromoplasts and give many fruits, flowers, and roots their bright colors [3]. For these reasons, assessing the presence and distribution of carotenoids at the subcellular level provides a keyhole through which to understand a variety of cellular processes. The use and utility of Raman imaging is addressed here toward this end.

The Raman effect involves an incoming photon being scattered inelastically by a polyatomic molecule. This phenomenon requires that energy is exchanged between the photon and molecule. Since energy levels in the molecule are discrete, the difference in energy between the incoming and scattered photon corresponds to a molecular transition, most typically a vibration [4]. Molecular vibrations are influenced by the composition of atoms that make up the molecule, the types of bonds that connect atoms, and molecular symmetry. Thus, Raman spectroscopy has been exploited as a highly sensitive analytical tool for the determination of the chemical identity of many molecules [5, 6]. The probability of observing the Raman effect is low, however, as the intensity of scattered radiation scales with the fourth power of incident light's frequency. The Raman effect is oftentimes considerably weaker as compared to the intrinsic fluorescence in cells and tissues, complicating and even prohibiting the detection of the Raman scatter in biologically relevant matrices.

However, if the frequency of incident light matches the energy of an electronic transition, an enhancement of the Raman signal may be observed. This is known as resonance Raman scattering (RRS), which can enhance the signal by several orders of magnitude over a "normal," non-resonance measurement. In fact, the resonance effect can increase the scattering cross section to exceed any off-resonance Raman scatter and, importantly, even rival intrinsic fluorescence from the species. RRS thus has profound analytical potential and has been comprehensively reviewed elsewhere for the general life science field [7] and specifically for applications in photosynthesis [8]. Here, its role in the assessment of carotenoids is examined explicitly.

The optical properties of carotenoids make them especially suitable for RRS. Carotenoids are π-electron-conjugated carbon-chain molecules consisting of alternating C—C single bonds and C=C double bonds. Individual carotenoids can be distinguished by the number of conjugated carbon double bonds, the number of attached methyl side groups, and the number and type of end groups. These properties result in many of the highly abundant carotenoid molecules (e.g., β-carotene, zeaxanthin, lycopene, and lutein) having distinct, yet broad (100 nm) absorption bands in the visible region of the spectrum. The absorption shifts to longer wavelengths as the effective conjugation length of the carotenoid increases. Fortuitously, these visible absorption bands overlap with the common laser wavelengths for Raman excitation. Thus, when excited under these conditions, carotenoids exhibit a very strong RRS response (enhancement factor of about five orders of magnitude relative to non-resonant Raman spectroscopy) and little to no fluorescence emission. Additionally, variations in absorption of the different carotenoids can be exploited by shifting the excitation wavelength (a technique also known as "tuning") to preferentially excite different carotenoid molecules. Owing to these amenities, RRS therefore enables the detection of carotenoids, even in complex biological systems, such as living photosynthetic cells and tissues.

The majority of carotenoids have linear structures resulting in a limited number of Raman-observable vibrations that are easily categorized into a distinct vibrational signature. There are three major Raman modes typically leveraged in the analysis of carotenoids [9, 10]. The ν_1 band (~1530 cm^{-1}) arises from the stretching vibrations of the conjugated C=C backbone. This band is sensitive to conjugation length and molecular conformation and therefore

http://dx.doi.org/10.5772/68101

is the most diagnostic for carotenoid identity. The ν_2 band (~1160 cm^{-1}) emerges from the stretching of the C—C vibrations coupled to C—CH_3 stretches or C—H in-plane bending. The ν_3 band (~1006 cm^{-1}) is attributed to CH_3 in-plane–rocking modes. Importantly, the vibrational modes are more or less insensitive to the molecular environment [11] meaning that a carotenoid found within the tissues of a plant may have a very similar spectrum to the same carotenoid dissolved in solvent. This observation is important when assigning carotenoid signatures *in situ*. **Figure 1** shows the resonance Raman spectra from six common carotenoids produced in photosynthetic organisms. The spectra are plotted in the order of increasing

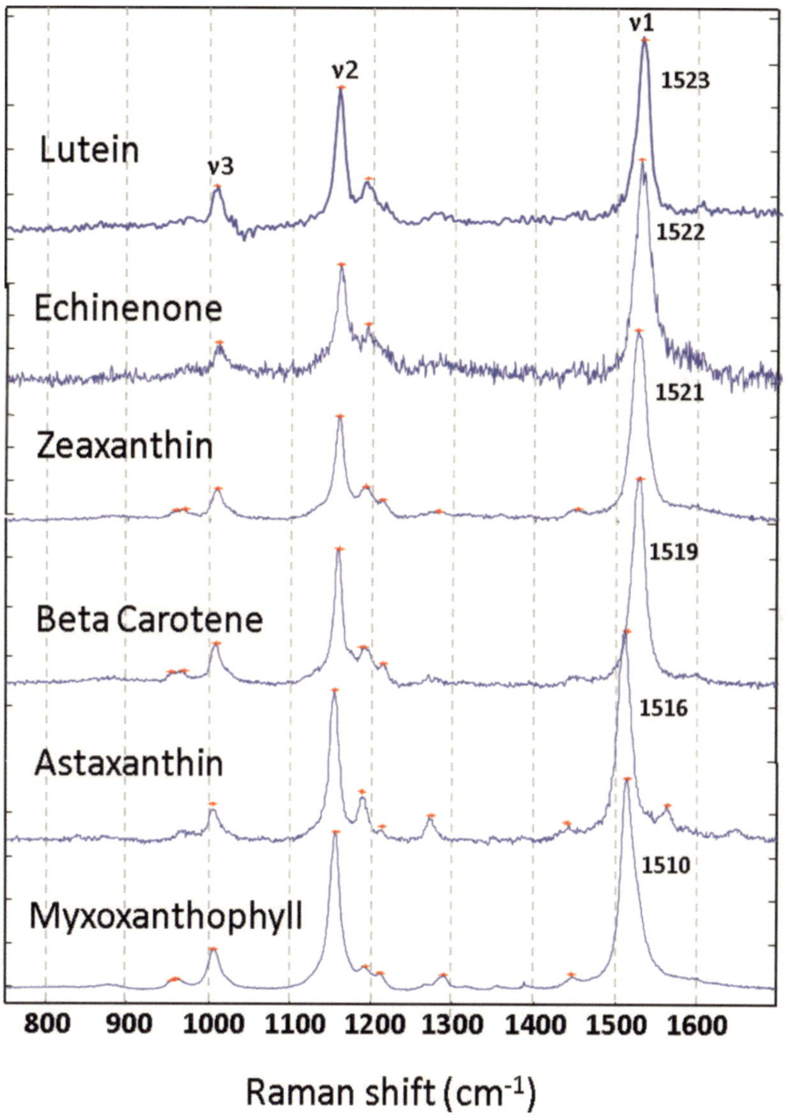

Figure 1. Resonance Raman spectra of six common carotenoids produced in photosynthetic organisms. Major Raman active vibrations are labeled. Spectra were obtained from carotenoids in powder form with the exception of lutein which was dissolved in methanol.

ν_1 vibration position (ranging from 1516 to 1524 cm^{-1}) and illustrate the ability to distinguish carotenoids by the position of this vibration.

In recent years, advances in detector sensitivity and acquisition speed have driven the expansion of Raman spectroscopy from a bulk analytical tool to a powerful method for mapping molecular vibrations in cells and tissues by the addition of a spatial dimension. Spatial resolution available with Raman microscopy is dependent on the wavelength of the excitation light and in theory is the same as other optical microscopies, such as fluorescence (e.g., ~250 nm lateral and ~500 nm axial, for blue/green excitation). The addition of spatial resolution can be accomplished in several different ways: confocal point-scanning Raman microscopy [12, 13], wide-field Raman imaging [14], or Raman line scanning [15, 16]. The relative advantages and disadvantages of each of these approaches have been reviewed elsewhere and will not be covered here [17]. In addition to the imaging methodology, spectral information can be obtained from either a single or a small number of bands through the use of discrete bandpass filters or tunable filters or in a hyperspectral fashion, where the entire Raman spectrum is dispersed via a prism or grating allowing for the simultaneous acquisition of hundreds of spectral bands. While there is typically a speed advantage associated with the filter-based acquisition approach, recent commercial systems utilizing electron-multiplied charge-coupled device (EM-CCD) detectors and cutting-edge research systems [18] are able to approach confocal fluorescence microscopy speeds and further improvements are expected.

Given the complexity associated with biological matrices such as live cells and tissues, hyperspectral Raman approaches are often advantageous because they can provide detailed spectral signatures for subsequent analysis of weak or overlapping features and background noise reduction using chemometric algorithms. These approaches are collectively referred in the literature as Raman spectroscopic imaging, or hyperspectral Raman microscopy. Additionally, confocal point scanning or pseudo-confocal line scanning can have advantages due to the inherent rejection of out-of-focus signal provided by the pinhole or entrance slit, respectively. A confocal, point-scanning, hyperspectral Raman microscope was utilized for the work presented in this chapter as it adequately rejects the intense pigment signal from outside the focal plane, while providing reasonable acquisition speeds for live cells.

Recent literature shows that Raman spectroscopic imaging is emerging as a key technology for single-cell analysis, including *in vivo* lipidomics [19], chemical composition of bacterial cells [20], lignin in plant cell walls [21], and metabolism in combination with stable isotope incorporation [22–24]. Researchers have also begun capitalizing on the specific benefits of resonance Raman spectroscopic imaging to localize carotenoids in individual cells and tissues. Pudney et al. demonstrated the localization of lutein, beta-carotene, and lycopene in several tomato varieties [25]. Zheng and coworkers used Raman imaging to investigate biofilm formation in Rhodococcus sp. SD-74, showing that biofilm production by this organism is correlated with increasing carotenoid concentrations, a finding not possible with traditional dye staining assays or electron microscopy [26]. Kilcrease et al. utilized a novel combination of confocal Raman microscopy with laser scanning confocal and scanning and transmission electron microscopy to demonstrate subcellular accumulation sites of various carotenoids in *Capsicum annuum* L. (Chile pepper) fruit. Their work discovered an unexpected relationship between carotenoid accumulation and chromoplast structure. Collins and coworkers utilized hyperspectral confocal Raman microscopy to study fundamental carotenoid biogenesis in *Haematococcus pluvialis* throughout the organism's life cycle [27]. Additionally, although not single cell work, Toomey et al. recently show the utility of hyperspectral confocal Raman microscopy for investigating carotenoid composition in single oil droplets within the avian retina [28].

While Raman spectroscopic imaging has potential for assessing carotenoid distributions in single cells and tissues for many applications in biomedicine and photosynthesis, there are some noteworthy limitations. First, not all carotenoids are enhanced to the same degree and some will not be enhanced at all. This differential enhancement can be advantageous, but it also can pose limitations for certain carotenoids. To some degree, the choice of laser excitation can target additional carotenoids. However, performing multiple Raman microscopy scans at two or more excitation wavelengths is prohibitive for live-cell dynamics and may require fixed samples depending on the acquisition times. Second, even with high spectral resolution instruments, multiple carotenoid species with highly overlapping, similar peaks can be difficult to identify. Additionally, even with resonance enhancement, Raman peaks may still be weak at *in vivo* and *in planta* relevant concentrations and be overwhelmed by background fluorescence depending on the sample matrix.

In addition to the hardware approaches listed above, the first two limitations are actively being addressed with the development of robust multivariate analysis tools [29–31]. For example, multivariate curve resolution (MCR) has been developed by several research groups for the analysis of hyperspectral confocal fluorescence and Raman image data sets [29, 32–34] finding success in complex multicomponent biological samples. It is therefore used in the analysis presented in this chapter. Lastly, while recent advances in fluorescence microscopy have extended the spatial resolution beyond the diffraction limit, the spatial resolution of Raman spectroscopic imaging is still in many cases orders of magnitude larger than the biological processes being investigated. Near-field approaches can provide higher spatial resolution, but have not found wide-scale success with living cells and tissues [35].

This chapter presents three separate applications of confocal Raman microscopy to assess carotenoid localization and relative abundance within living cells. The green algae *H. pluvialis* is presented to highlight the ability to discern highly similar carotenoid structures in algal cells. Maize (*Zea mays*) protoplasts demonstrate the localization of carotenoids even amidst a high background of chlorophyll pigment. Finally, *Synechocystis* sp. PCC6803, a model cyanobacterium, is presented to illustrate carotenoid localization even in very small microbes. Together, these applications demonstrate the potential of resonance Raman microscopy and imaging for the analysis of carotenoid localization and abundance at the single-cell and subcellular levels in photosynthetic organisms.

2. Materials and methods

2.1. Cell culture and sample prep

H. pluvialis cells were obtained from the Hu lab cultured as described previously to produce cultures at various stages of the life cycle (flagellated, nonmotile palmelloid, and aplanospores) [27]. Cells suspended in growth media were directly loaded to a slide and covered with a glass coverslip. In this arrangement, cells remained hydrated during the duration of the imaging.

Synechocystis 6803 cells were obtained from the Pakrasi lab and cultured in BG-11 media (UTEX recipe) in baffled flasks in a temperature-controlled incubator under continuous light (~30 μmol m^{-2} s^{-1}, cool white fluorescent lights) with shaking (60 rpm). Prior to imaging, a 4-μl sample was directly pipetted onto a BG-11 agar-coated slide and covered with a glass coverslip. Cells were immediately imaged.

Etiolated maize protoplasts were isolated from *Z. mays* var. B73 leaves as described previously [36]. Released protoplasts were collected, washed, and shipped in a buffer solution overnight for Raman imaging. Upon arrival, protoplasts were concentrated by centrifugation at 500 × g and removal of all but 100 μl of the supernatant. Concentrated sample (25 μl) was loaded into an *in situ* frame (Gene Frame, ThermoFisher) previously adhered to a gridded microscope slide (Lovins Micro-Slide Field Finder, Electron Microscopy Sciences) and covered with a glass coverslip. The gridded slide was used to perform correlated fluorescence microscopy for a separate experiment. Protoplasts were immediately imaged.

2.2. Confocal Raman microscopy

Raman images were acquired with a WiTec Alpha300R system equipped with a WiTec UHTS spectrometer utilizing a 600-l/mm grating and an Andor back-illuminated electron-multiplying charge-coupled device (EMCCD). Light was incident at 532 nm and focused using a 50×/0.55 NA objective (*Synechocystis* 6803), 20×/0.45 NA objective (maize protoplasts), and a 100×/0.9 NA objective (*H. pluvialis*). Laser powers were chosen such that the spectral character was nearly invariant with time and did not cause a visible change in the sample during collection. In most cases, power incident on the surface was held to less than 1 mW. The acquisition time per spectrum was chosen to provide adequate signal to noise to perform the multivariate analysis and varied for the different samples analyzed ranging from 4 to 10 ms per spectrum. The spectral response was 3 cm^{-1} per pixel, which through fitting resulted in the ability to specify peak position to an absolute accuracy of ±1 cm^{-1}. All measurements were performed in an unpolarized back-scattering arrangement. Images were acquired by scanning the sample using a piezo-stage possessing a lateral resolution of <5 nm. Raman-scattered light is collected continuously across the sample such that every spectrum is the average convolution of the beam intensity and sample response over the distance defining an individual voxel. Images were collected by acquiring a spectrum every 333 nm (*Synechocystis* 6803 and *H. pluvialis*) or 500 nm (maize protoplasts). Raman shifts were calibrated using the well-established positions of silicon, 6H-SiC, graphite, and 4-acetameniphol.

2.3. Image data analysis

All spectral image analysis was performed in Matlab 2012 or 2015 (Mathworks) equipped with the statistics and machine learning, signal processing, image processing, and curve fitting toolboxes leveraging in-house written software, functions, and scripts. Hyperspectral confocal Raman images were preprocessed to remove cosmic spikes [37]. When applicable, images from the same sample were compiled into a composite image data set. The use of composite images, rather than analyzing every image independently, increases the number of pixels for the MCR analysis and thus serves to improve spectral signature identification by adding additional variance [38]. The spectral region was trimmed to exclude the excitation laser line but still include about 5–10 pixels of the spectrum where the signal was blocked by the Rayleigh filter. This technique assists in the analysis by providing a "zero-signal region" for assessing baseline contributions and is discussed in detail by Jones et al. [38]. In most cases, an image mask was created that excluded pixels outside the area of the cells from the analysis as they contain predominantly background signal.

Principal components analysis (PCA) was performed on the composite image and the Scree plot was inspected to determine the number of independent components present in the image data set (measured as being before the bend in the elbow of the Scree plot [29]). Multivariate curve resolution was then performed using a constrained alternating least-squares algorithm and employing robust constraints for equality (offset/baseline) and non-negativity (all true spectral components) to develop a spectral model that described the spectral variance within the data set. A PCA analysis of the residuals was used to confirm the appropriateness of the spectral model for describing the data and identify any unmodeled spectral signatures. The multivariate curve resolution algorithm [39–43] and specific approaches for success with biological images have been described in detail elsewhere [29, 38]. The details of the spectral models developed are presented during the results and discussion for each application in this chapter. The MCR-identified spectra were then used in a classical least-squares (CLS) analysis to predict the concentrations of each spectral component in each image pixel. Lastly, the resulting concentration maps were exported as 16-bit grayscale tiffs such that they could be subject to traditional image analysis. Color image overlays and simple image cropping and scaling for visualization purposes was performed in Fiji [44].

3. Results and discussion

3.1. Resolving different carotenoids in living cells: *H. pluvialis*

H. pluvialis is a freshwater green microalga that can synthesize and accumulate astaxanthin, a high-value nutraceutical, under stress conditions such as nutrient deprivation. Though some aspects of carotenogenesis leading to astaxanthin production are well understood, spatial-temporal details of key carotenoids and enzymes have yet to be fully elucidated. In previous work, Collins and coworkers used confocal Raman spectroscopic imaging to provide *in vivo* resolution and localization of beta-carotene and astaxanthin, along with chlorophyll throughout the life cycle of *H. pluvialis* cells [27]. While this work will not be reproduced in detail, here, some key elements will be presented to illustrate the potential of the technique for carotenoids localization in living cells. **Figure 2** demonstrates the advantage of performing MCR analysis on the Raman spectroscopic image data as compared to simplistic band integration. While the image produced by integrating the area under the ν_1 carotenoid vibration provides a highly resolved view of carotenoid location within the cell, the image is a superposition of two different carotenoids found in the cell. By contrast, MCR analysis resulted in a four-component model consisting of two carotenoid spectra (astaxanthin and beta-carotene), chlorophyll, and an autofluorescence component. (Note: A background component was also part of the model, but is omitted here for simplicity.) MCR analysis also generates a corresponding concentration map (i.e., image) for each of these species. This results in an exquisite degree of chemical resolution that allows for quantitative assessment via traditional image analysis methods. In **Figure 2D**, the images for astaxanthin, beta-carotene, and chlorophyll are pseudocolored and overlaid for qualitative comparison. Importantly, in this example, the MCR analysis reveals that astaxanthin is located only in the cytosol. Beta-carotene, meanwhile, is both co-located with astaxanthin in the cytosol, as well as with the chlorophyll in the chloroplast. Together, these Raman images provide spatial-temporal details of astaxanthin production and accumulation in this organism.

3.2. Carotenoids in high-chlorophyll backgrounds: maize protoplasts

Similar to algae, plant carotenoids have roles in light harvesting and protection against light and heat stress. Metabolic engineering of carotenoids in plants has the potential to create varieties exhibiting increased adaptation to climate change as well as additional nutritional value. To develop plant cultivars with these enhanced properties, it is critical to understand the detailed spatial-temporal arrangement of both carotenoids and chlorophyll *in situ*. While chlorophyll is straightforward to detect with a standard confocal fluorescence microscope, carotenoids have little to no detectable fluorescence signal and are not amenable to exogenous labeling.

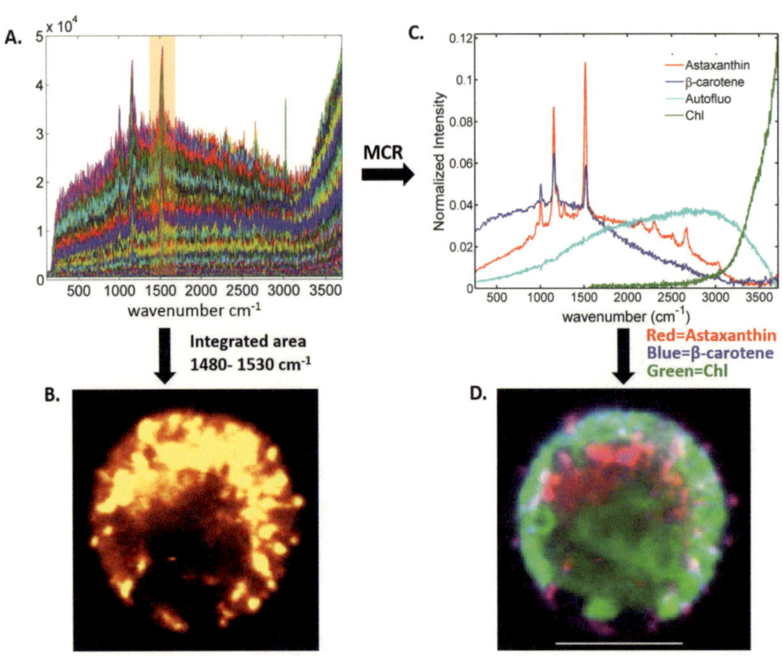

Figure 2. Raman-spectroscopic imaging of *H. pluvialis*: the advantage of MCR analysis. (A) Raw spectra from Raman-spectroscopic image of *H. pluvialis*. (B) Integrated Raman image created by traditional integration over the area of the v_1 band of the carotenoid (1480–1530 cm^{-1}) for the image data in A. C. MCR identified spectral component from image data in A. RGB image created by overlaying the independent component maps generated in the MCR analysis. Scale bar = *10 μm*. (note: **Figure 2** is best viewed in color. Please refer to the online version of this chapter for the color version of **Figure 2**).

Confocal Raman spectroscopic imaging is capable of detecting carotenoids without the use of any exogenous labels and does so by detecting the resonance Raman signal of carotenoids. This signal can be observed also within the background of other pigments, such as the chlorophyll precursor, protochlorophyllide, which is found in cells of etiolated tissue [45] such as plant protoplasts from etiolated maize (**Figure 3**). Plant protoplasts are plant cells that have their cell wall removed through enzyme treatment and thus are an excellent experimental system for plant biologists because they improve the ease of introducing new genes.

Figure 3 highlights the spatial location and levels of carotenoids and protochlorophyllide in plastids (immature chloroplasts) of etiolated maize protoplasts as observed by confocal Raman microscopy. Previous high-performance liquid chromatography data have shown that the dominant carotenoid in maize-etiolated protoplasts is lutein with a minor amount of violaxanthin (data not shown). The peak positions (ν_1 = 1525 cm^{-1} and ν_2 = 1157 cm^{-1}) in the isolated Raman spectrum shown in **Figure 3C** are consistent with the purified lutein spectrum in **Figure 1** and by Collins et al. [27]. The slight deviation is likely a result of the violaxanthin contribution as the ν_1 vibration of violaxanthin is typically 1528–1530 cm^{-1} [46]. For comparison, Panels A and B show the correlated bright-field and confocal fluorescence chlorophyll images, respectively, for the cell shown in **Figure 3D**. Individual component images are shown in grayscale as well as the red-green merged image (please refer to the online version for color figure). Green corresponds to the carotenoid and red corresponds to the protochlorophyllide pigments in these figures. Localization is confined to the plastids (bright foci ~1 μm in size within the protoplast). Interestingly, a high degree of variation in carotenoid and chlorophyll composition within individual plastids is observed. An example of this heterogeneity is indicated by the yellow arrows in the merged image of **Figure 3D**. These arrows highlight plastids that have predominantly carotenoid (green), predominantly protochlorophyllide (red), or roughly equal amounts of both (yellowish-orange). These differences in relative abundance could arise from differences in the developmental phase or carotenoid abundance. The observed heterogeneities raise important scientific questions about carotenoid biogenesis that Raman-spectroscopic imaging is well poised to answer in future research.

3.3. Approaching the limits of spatial resolution: *Synechocystis* 6803

Cyanobacteria and closely related organisms are thought to be the evolutionary ancestors to chloroplasts found in plants. Cyanobacteria perform photosynthesis to convert light energy to chemical energy and play key roles in the ecology of the earth. Light is harvested by photosynthetic antennae, which are pigment-protein complexes composed of pigments with varying absorption and emission properties in order to "funnel" the energy to the reaction center where it is converted to chemical energy [47]. Carotenoids are integrated into the membrane-associated photosynthetic antennae complexes as well as the photosystems of all cyanobacteria and fundamentally coupled to the processes of light harvesting and photosynthesis. Unfortunately, carotenoids in cyanobacteria are often quite difficult to localize *in situ* because of the very small size of many of the cyanobacteria (1.5–4 μm for the most species) and the co-localization in complexes with highly fluorescent light-harvesting pigments, like phycobilins and chlorophylls. Previous work with hyperspectral confocal fluorescence microscopy has demonstrated sensitivity for localization carotenoids via resonance Raman signatures in

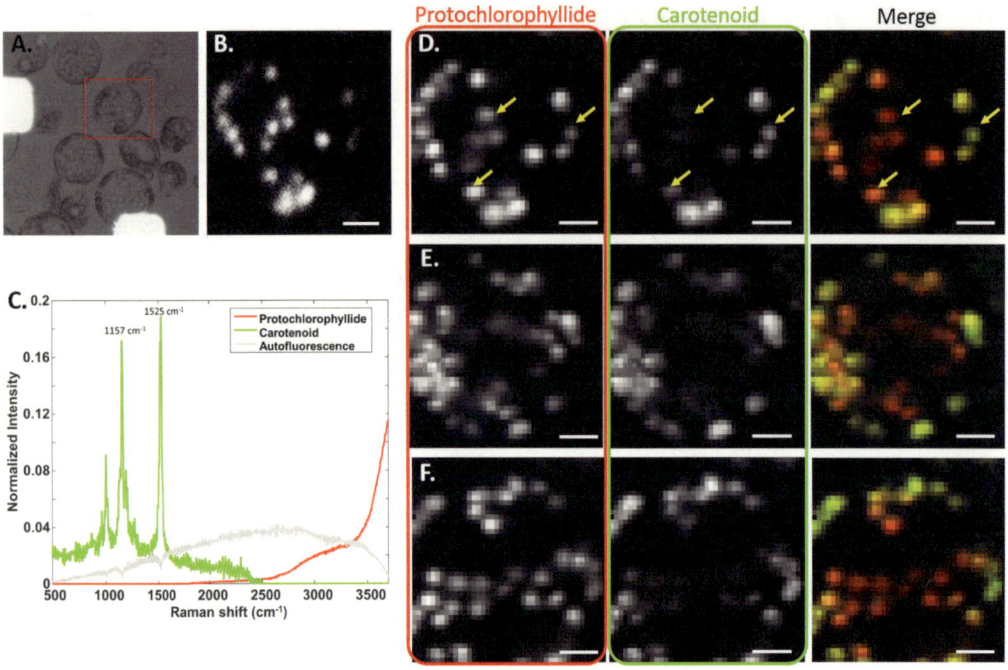

Figure 3. Confocal Raman-spectroscopic imaging of carotenoids and protochlorophyllide in etiolated maize protoplasts. (A) Bright-field image of several protoplasts with red square highlighting single protoplast imaged in B and D. (B) Confocal fluorescence image of cell highlighted in A (chlorophyll emission channel). (C) Spectral components of MCR model from the analysis of confocal Raman image of maize protoplasts. (D) Left panel: protochlorophyllide concentration map. Center panel: carotenoid concentration map. Right panel: Merged red and green color overlay, where the protochlorophyllide image is assigned to the red channel and carotenoid image is assigned to the green channel. Arrows highlight examples of plastid heterogeneity. (E) and (F). Two additional protoplasts. Image panels for E and F are the same as in D. Scale bars = 5 μm. (note: **Figure 3** is best viewed in color. Please refer to the online version of this chapter for the color version of **Figure 3**).

Synechocystis 6803 [48]; however, the spectral resolution of such a system is not sufficient to resolve different carotenoids.

Synechocystis 6803 has emerged as a model cyanobacterium due to the ease of genetic manipulation. Although *Synechocystis* 6803 cells are quite small (1.5–2 μm), Raman spectroscopic imaging can provide information about subcellular carotenoid localization. **Figure 4** shows the results of Raman spectroscopic imaging of wild-type *Synechocystis* 6803 cells. The carotenoid spectrum isolated in **Figure 4A** has ν_1 vibration appearing at 1515 cm^{-1} and ν_2 vibration at 1155 cm^{-1}, identifying that carotenoid most likely corresponds to beta-carotene, one of the major carotenoid species in *Synechocystis* 6803. The phycobilin spectrum peak is centered at ~3250 cm^{-1} (equivalent to 643.2 nm) and thus corresponds to phycocyanin, the primary pigment found in the antenna structure. The fluorescence emission of phycocyanin is ~645 nm. While subcellular details are somewhat limited given the small number of pixels in each cell, cell-to-cell differences in localization patterns are clearly evident. For example, in **Figure 4**, the carotenoids in the dividing cell on the left are concentrated into one small foci on each cell, whereas in the other, two cells have carotenoids localized more uniformly through the thylakoid membranes that house the light-harvesting complexes, similar to the phycocyanin.

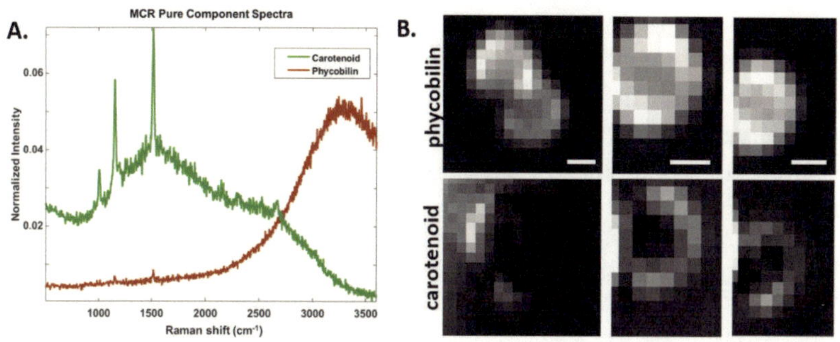

Figure 4. Results from confocal Raman-spectroscopic imaging of wild-type *Synechocystis* 6803 cells. (A) MCR identified pure component spectra. A fourth component consisting of a tail at ~3500 cm^{-1} was isolated as well and corresponds to chlorophyll. It was omitted for simplicity. (B) MCR concentration maps indicating carotenoid (bottom row) and phycobilin (top row) abundance for three different cells. Image intensities have been independently scaled between for maximum visibility; however, all intensity scales are within 20% of each other. A concentration map for the offset component is not shown. Scale bar = 1 μm.

Cell-to-cell heterogeneity and population dynamics are important to understand cyanobacterial adaptation to changes in the abiotic or biotic environment and could be addressed with Raman spectroscopic imaging in the future. It is also important to note that the images in **Figure 4** were collected under slightly less than diffraction-limited conditions. Improvements in spatial resolution are possible with a more optimal arrangement.

4. Conclusions

Localizing carotenoids in living cells and tissues is challenging due to the complex biological matrices of the living cells and intense sometimes colocated interfering fluorescence. Raman spectroscopic imaging coupled with multivariate curve resolution analysis provides the necessary spatial and chemical resolution to identify highly similar carotenoids even in the midst of highly overlapping, strong chlorophyll emission and complex backgrounds. Three examples were presented that highlight different advantages of the methodology for investigating photosynthetic organisms. Recent technological advances in detector speed and sensitivity will most likely catalyze future investigations of carotenoid biogenesis in single cells, including population dynamics and response to changing environmental conditions, by facilitating time-course studies that were previously prohibitive given the long scan times required for Raman spectroscopic imaging. These capabilities are anticipated to impact a variety of research areas including carbon partitioning and utilization, microbial ecology, and crop analytics.

Acknowledgements

The authors are grateful to the following people for their assistance with the research presented in this chapter: Anthony McDonald for assistance with the collection of Raman spectroscopic image data; Meghan Dailey for culturing the *Synechocystis* 6803; Stephen Anthony for the current implementation of the MCR analysis software used to deconvolve spectral components; Howland Jones, Mark Van Benthem, David Melgaard, Mike Keenan, and David Haaland for original development of the MCR algorithm and software; Wim Vermaas, Arizona State University, for the gift of the purified carotenoid standards, Himadri Pakrasi, Washington University, St. Louis, for the gift of *Synechocystis sp.* PCC 6803; Qiang Hu for the gift of the *H. pluvialis.*

This research was primarily supported as part of the Photosynthetic Antenna Research Center (PARC), an Energy Frontier Research Center funded by the U.S. Department of Energy (DOE), Office of Science, Basic Energy Sciences (BES), under Award # DE-SC 0001035 (Maize protoplasts imaging, Synechocystis imaging, writing), by Sandia National Laboratories' Laboratory Directed Research and Development (LDRD) Program under Award # 141528 (*H. pluvialis* imaging), and by funding from the National Institutes of Health (GM081160) (to ETW) for research on carotenoids in maize protoplasts. Sandia National Laboratories is a multimission laboratory managed and operated by Sandia Corporation, a wholly owned subsidiary of Lockheed Martin Corporation, for the U.S. Department of Energy's National Nuclear Security Administration under contract DE-AC04-94AL85000.

Author details

Jerilyn A. Timlin[1]*, Aaron M. Collins[2], Thomas A. Beechem[1], Maria Shumskaya[3,5] and Eleanore T. Wurtzel[3,4]

*Address all correspondence to: jatimli@sandia.gov

1 Sandia National Laboratories, Albuquerque, NM, USA

2 Southern New Hampshire University, Manchester, NH, USA

3 Department of Biological Sciences, Lehman College, The City University of New York (CUNY), Bronx, New York, NY, USA

4 The Graduate School and University Center-CUNY, New York, NY, USA

5 Department of Biology, School of Natural Sciences, Kean University, Union, NJ, USA

References

[1] Frank HA, Cogdell RJ. Carotenoids in photosynthesis. Photochem Photobiol. 1996;**63**: 257–64. DOI: 10.1111/j.1751-1097.1996.tb03022.x.

[2] Demmig-Adams B, Adams Iii WW. The role of xanthophyll cycle carotenoids in the protection of photosynthesis. Trends Plant Sci. 1996;**1**:21–6. DOI: 10.1016/S1360-1385(96)80019-7.

[3] Bartley GE, Scolnik PA. Plant carotenoids: pigments for photoprotection, visual attraction, and human health. The Plant Cell. 1995;**7**:1027–38.

[4] Lewis IR, Edwards HGM, editors. Handbook of Raman Spectroscopy. New York: Marcel Dekker, Inc.; 2001.

[5] Vankeirsbilck T, Vercauteren A, Baeyens W, Van der Weken G, Verpoort F, Vergote G, et al. Applications of Raman spectroscopy in pharmaceutical analysis. TrAC Trends in Analytical Chemistry. 2002;**21**:869–77. DOI: 10.1016/S0165-9936(02)01208-6.

[6] Clegg IM, Everall NJ, King B, Melvin H, Norton C. On-line analysis using Raman spectroscopy for process control during the manufacture of titanium dioxide. Appl Spectrosc. 2001;**55**:1138–50. DOI: 10.1366/0003702011953388.

[7] Efremov EV, Ariese F, Gooijer C. Achievements in resonance Raman spectroscopy: review of a technique with a distinct analytical chemistry potential. Anal Chim Acta. 2008;**606**:119–34. DOI: 10.1016/j.aca.2007.11.006.

[8] Robert B. Resonance Raman spectroscopy. Photosynth Res. 2009;**101**:147–55. DOI: 10.1007/s11120-009-9440-4.

[9] Merlin JC. Resonance Raman spectroscopy of carotenoids and carotenoid-containing systems. Pure Appl Chem. 1985;**57**:785–92.

[10] Rimai L, Heyde ME, Gill D. Vibrational spectra of some carotenoids and related linear polyenes. Raman spectroscopic study. J Am Chem Soc. 1973;**95**:4493–501. DOI: 10.1021/ja00795a005.

[11] Gill D, Kilponen RG, Rimai L. Resonance Raman scattering of laser radiation by vibrational modes of carotenoid pigment molecules in intact plant tissues. Nature. 1970;**227**:743–4.

[12] Klein K, Gigler Alexander M, Aschenbrenner T, Monetti R, Bunk W, Jamitzky F, et al. Label-free live-cell imaging with confocal Raman microscopy. Biophys J. 2012;**102**:360–8. DOI: 10.1016/j.bpj.2011.12.027.

[13] Dieing T, Hollricher O, Toporski J, editors. Confocal Raman Microscopy: Springer-Verlag Berlin Heidelberg Springer; 2011.

[14] Morris HR, Hoyt CC, Miller P, Treado PJ. Liquid crystal tunable filter Raman chemical imaging. Appl Spectrosc. 1996;**50**:805–11.

[15] Timlin JA, Carden A, Morris MD. Chemical microstructure of cortical bone probed by Raman transects. Appl Spectrosc. 1999;**53**:1429–135.

[16] Christensen KA, Morris MD. Hyperspectral Raman microscopic imaging using Powell lens line illumination. Appl Spectrosc. 1998;**52**:1145–7.

[17] Schlucker S, Schaeberle MD, Huffman SW, Levin IW. Raman microspectroscopy: A comparison of point, line, and wide-field imaging methodologies. Anal Chem. 2003;**75**:4312–8.

[18] Kong L, Navas-Moreno M, Chan JW. Fast confocal Raman imaging using a 2-D multifocal array for parallel hyperspectral detection. Anal Chem. 2016;**88**:1281–5. DOI: 10.1021/acs.analchem.5b03707.

[19] Wu H, Volponi JV, Oliver AE, Parikh AN, Simmons BA, Singh S. *In vivo* lipidomics using single-cell Raman spectroscopy. Proc Natl Acad Sci. 2011;**108**:3809–14. DOI: 10.1073/pnas.1009043108.

[20] Schuster KC, Urlaub E, Gapes JR. Single-cell analysis of bacteria by Raman microscopy: spectral information on the chemical composition of cells and on the heterogeneity in a culture. J Microbiol Meth. 2000;**42**:29–38.

[21] Gierlinger N, Keplinger T, Harrington M. Imaging of plant cell walls by confocal Raman microscopy. Nat Protocols. 2012;**7**:1694–708. DOI: 10.1038/nprot.2012.092.

[22] Huang WE, Griffiths RI, Thompson IP, Bailey MJ, Whiteley AS. Raman microscopic analysis of single microbial cells. Anal Chem. 2004;**76**:4452–8. DOI: 10.1021/ac049753k.

[23] Huang WE, Stoecker K, Griffiths R, Newbold L, Daims H, Whiteley AS, et al. Raman-FISH: combining stable-isotope Raman spectroscopy and fluorescence in situ hybridization for the single cell analysis of identity and function. Environ Microbiol. 2007;**9**:1878–89. DOI: 10.1111/j.1462-2920.2007.01352.x.

[24] Matthäus C, Krafft C, Dietzek B, Brehm BR, Lorkowski S, Popp J. Noninvasive imaging of intracellular lipid metabolism in macrophages by Raman microscopy in combination with stable isotopic labeling. Anal Chem. 2012;**84**:8549–56. DOI: 10.1021/ac3012347.

[25] Pudney PDA, Gambelli L, Gidley MJ. Confocal Raman microspectroscopic study of the molecular status of carotenoids in tomato fruits and foods. Appl Spectrosc. 2011;**65**:127–34. DOI: 10.1366/10-06121.

[26] Zheng Y-T, Toyofuku M, Nomura N, Shigeto S. Correlation of carotenoid accumulation with aggregation and biofilm development in Rhodococcus sp. SD-74. Anal Chem. 2013;**85**:7295–301. DOI: 10.1021/ac401188f.

[27] Collins AM, Jones HDT, Han D, Hu Q, Beechem TE, Timlin JA. Carotenoid distribution in living cells of *Haematococcus pluvialis* (Chlorophyceae). PLoS One. 2011;**6**:e24302.

[28] Toomey MB, Collins AM, Frederiksen R, Cornwall MC, Timlin JA, Corbo JC. A complex carotenoid palette tunes avian colour vision. J Royal Soc Interf. 2015;**12**. DOI: 10.1098/rsif.2015.0563.

[29] Haaland DM, Jones HDT, Timlin JA. Experimental and Data Analytical Approaches to Automating Multivariate Curve Resolution in the Analysis of Hyperspectral Images. In: Ruckebusch C, editor. Resolving Spectral Mixtures. Data Handling in Science and Technology. 30. Amsterdam: Elsevier; 2016. pp. 381–406.

[30] Zhang J, O'Connor A, Turner II JF. Cosine histogram analysis for spectral image data classification. Appl Spectrosc. 2004;**58**:1318–24.

[31] Widjaja E, Crane N, Chen T-c, Morris MD, Ignelzi Jr MA, McCreadie BR. Band-target entropy minimization (BTEM) applied to hyperspectral Raman image data. Appl Spectrosc. 2003;**57**:1353–62.

[32] Felten J, Hall H, Jaumot J, Tauler R, de Juan A, Gorzsás A. Vibrational spectroscopic image analysis of biological material using multivariate curve resolution–alternating least squares (MCR-ALS). Nat Protocols. 2015;**10**:217–40. DOI: 10.1038/nprot.2015.008.

[33] de Juan A, Tauler R. Chemometrics applied to unravel multicomponent processes and mixtures: Revisiting latest trends in multivariate resolution. Anal Chim Acta. 2003;**500**:195–210. DOI: 10.1016/S0003-2670(03)00724-4.

[34] Schoonover JR, Marx R, Zhang SL. Multivariate curve resolution in the analysis of vibrational spectroscopy data files. Appl Spectrosc. 2003;**57**:154A–70A.

[35] Webster S, Batchelder DN, Smith DA. Submicron resolution measurement of stress in silicon by near-field Raman spectroscopy. Appl Phys Lett. 1998;**72**:1478–80. DOI: 10.1063/1.120598.

[36] Shumskaya M, Bradbury LMT, Monaco RR, Wurtzel ET. Plastid localization of the key carotenoid enzyme phytoene synthase is altered by isozyme, allelic variation, and activity. The Plant Cell. 2012;**24**:3725–41. DOI: 10.1105/tpc.112.104174.

[37] Anthony SM, Timlin JA. Removing cosmic spikes using a hyperspectral upper-bound spectrum method. Appl Spectrosc. 2016:0003702816668528. DOI: 10.1177/0003702816668528.

[38] Jones HDT, Haaland DM, Sinclair MB, Melgaard DK, Collins AM, Timlin JA. Preprocessing strategies to improve MCR analyses of hyperspectral images. J Chemomet Intell Lab Syst. 2012;**117**:149–58. DOI: 10.1016/j.chemolab.2012.01.011.

[39] Van Benthem MH, Keenan MR, inventors; Sandia Corporation, assignee. Fast combinatorial algorithm for the solution of linearly constrained least squares problems. US2008.

[40] Van Benthem MH, Keenan MR. Fast algorithm for the solution of large scale non-negativity constrained least squares problems. J Chemom. 2004;**18**:441–50.

[41] Ohlhausen JA, Keenan MR, Kotula PG, Peebles DE. Multivariate statistical analysis of time-of-flight secondary ion mass spectrometry images using AXSIA. Appl Surf Sci. 2004;**231-232**:230–4.

[42] Kotula PG, Keenan MR, Michael JR. Automated analysis of SEM X-Ray spectral images: a powerful new microanalysis tool. Microsc Microanal. 2003;**9**:1–17.

[43] Van Benthem MH, Keenan MR, Haaland DM. Application of equality constraints on variables during alternating least squares procedures. J Chemom. 2002;**16**:613–22.

[44] Schindelin J, Arganda-Carreras I, Frise E, Kaynig V, Longair M, Pietzsch T, et al. Fiji: an open-source platform for biological-image analysis. Nat Meth. 2012;**9**:676–82. DOI: 10.1038/nmeth.2019.

[45] Robertson DS, Bachmann MD, Anderson IC. Role of carotenoids in protecting chlorophyll from photodestruction—II. Studies on the effect of four modifiers of the albino cl1 mutant of maize. Photochem Photobiol. 1966;**5**:797–805. DOI: 10.1111/j.1751-1097.1966.tb05775.x.

[46] Ruban AV, Pascal A, Lee PJ, Robert B, Horton P. Molecular configuration of xanthophyll cycle carotenoids in photosystem II antenna complexes. J Biol Chem. 2002;**277**:42937–42. DOI: 10.1074/jbc.M207823200.

[47] Blankenship RE. Molecular Mechanisms of Photosynthesis: Wiley-Blackwell; Oxford 2013. 336 p.

[48] Vermaas WFJ, Timlin JA, Jones HDT, Sinclair MB, Nieman LT, Hamad S, et al. *In vivo* hyperspectral confocal fluorescence imaging to determine pigment localization and distribution in cyanobacterial cells. Proc Natl Acad Sci. 2008;**105**:4050–5.

Chapter 4

The Biochemistry and Antioxidant Properties of Carotenoids

Oguz Merhan

Additional information is available at the end of the chapter

http://dx.doi.org/10.5772/67592

Abstract

Carotenoids are one of the most widespread pigment groups distributed in nature, and more than 700 natural carotenoids have been described so far, and new carotenoids are introduced each year. Carotenoids are derived from 4 terpenes, including totally 40 carbon atoms. Carotenoids are naturally synthesized by cyanobacteria, algae, plants, some fungi, and some bacteria, but not made by mammals. Lately, the beneficial properties of α-carotene, β-carotene, γ-carotene, lycopene, phytoene, phytofluene, lutein, zeaxanthin, β-cryptoxanthin, astaxanthin, and fucoxanthin carotenoids in prevention of various diseases, such as tumor formation, cardiovascular, and vision, have been documented due to their roles as antioxidants, activation in certain gene expression associated with cell-to-cell communication, provitamin A activity, modulation of lipoxygenase activity, and immune response. In this chapter, in addition to biochemical properties of carotenoids, how the structure of these molecules influences the oxidative stress in health and reducing the risk of formation of various diseases will be described.

Keywords: carotenoids, antioxidants, biochemistry, health, toxicity

1. Introduction

Overproduction of reactive oxygen and nitrogen species, such as superoxide and peroxide radicals, plays important roles in the formation of breast, cervical, ovarian, and colorectal cancer in addition to some other malignancies in the cardiovascular system and eye. The antioxidant property of carotenoids may result from its double carbon-carbon bonds interacting with each other via conjugation and causing electrons in the molecule to move freely across these areas of the molecule. Carotenoid intake from food sources reduces the risk of breast, lung, head and neck, cervical, ovarian, colorectal, and prostate cancers and cardiovascular or eye diseases due to their roles as antioxidants. It has been reported that carotenoids can directly interact with some free radical species.

2. The biochemistry and metabolism of main carotenoids

2.1. Terpenes

Terpenes are structures derived from isoprene chains. The chemical formula of the isoprene molecule is $CH_2{=}C(CH_3){-}CH{=}CH_2$ (2-methyl-1,3-butadiene) [1, 2]. There are two double bonds in the molecule and these bonds are conjugated (**Figure 1**). When 5-carbon containing isoprene molecules are polymerized, compounds called terpenes are formed [3]. This group includes biologically very important molecules. Some among these are lycopene, β-carotene, vitamin A, and squalene [4].

Terpenes are biosynthetically derived from isoprene units, the chemical formula of this unit is C_5H_8, and the basic molecular formulas of terpenes are multiplied, that is, $(C_5H_8)_n$ (n is the number of joined isoprene units). The isoprene units can be connected from head to tail to form straight chains or rings. The isoprene unit is a building brick that is widely used in nature [5, 6].

Terpenes can be classified according to the number of isoprene units used, such as hemiterpenes (prenol and isovaleric acid); one isoprene unit (5C), monoterpenes (geraniol and limonene); two isoprene units (10C), sesquiterpenes (farnesol); three isoprene units (15C), diterpenes (quinagolides, sembren, and taxadiene); four isoprene units (20C), sesterterpenes; five isoprene units (25C), triterpenes (squalene); six isoprene units (30C), tetraterpenes (carotenoids); and eight isoprene units (40C), politerpenes, which include great number of isoprene units (natural rubber) [6–9].

Monoterpenes, sesquiterpenes, diterpenes, and sesterterpenes are formed by head-to-tail association of isoprene units, and triterpenes and tetraterpenes (carotenoids) are formed by head-to-head association [9].

The most important group of terpenes is carotenoids with the $C_{40}H_{64}$ molecular formula which is formed by a tetraterpene containing eight isoprene units [8]. Carotenoids are terpene

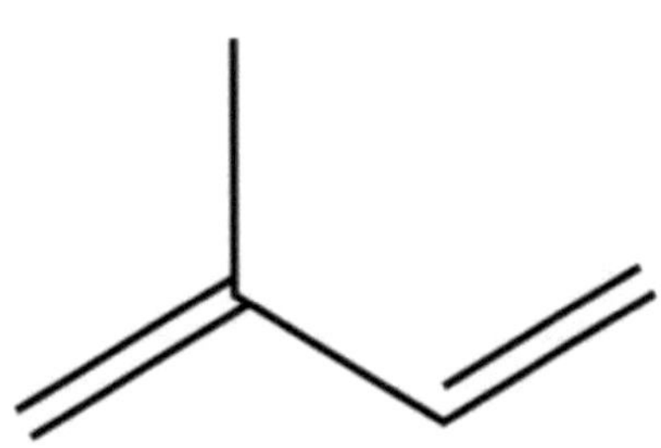

Figure 1. Chemical formula for isoprene.

http://dx.doi.org/10.5772/67592

group materials which are formed by binding of 5-carbon-containing isoprene molecules, and they have straight chain structure formed by the condensation of isoprene molecules [1]. Carotenoids which colors vary from light yellow to dark red stemming from the double bonds are soluble in organic solvents and oils like other lipids [10].

2.2. Biosynthesis of terpenes and carotenoids

Carotenoids which are tetraterpenes and biological pigments including eight isoprene units are found in plants and some other photosynthetic microorganisms (naturally occurring cyanobacteria, algae, plants, some fungi, and some bacteria) [11].

Acetyl CoA, which is synthesized directly from free acetate as well as the oxidation of sugars and condensation of pyruvic acid or fatty acids, is used as a precursor in the synthesis of mevalonic acid in addition to synthesis of many natural compounds [9, 12]. First, two acetyl CoA molecules enter the reaction to give acetoacetyl-CoA, and then β-hydroxy-β-methylglutaryl-CoA (HMG-CoA) is obtained via HMG-CoA synthase with another acetyl CoA. The reduction of the ester group in HMG-CoA by NADPH and HMG-CoA reductase (the pathway's rate limiting enzyme) results in mevalonic acid formation. Mevalonic acid later forms mevalonate 5-phosphate using a total of two molecules of ATP by mevalonate kinase, and then mevalonate 5-phosphate forms mevalonate 5-diphosphate by phosphomevalonate kinase (mevalonate phosphate kinase). Isopentenyl diphosphate (IPP) building blocks from decarboxylation of mevalonate diphosphate via mevalonate diphosphate carboxylase produce isoprene chains. IPP:dimethylallyl-PP isomerase converts IPP to dimethylallyl diphosphate (DMAPP) which is an acceptor by successive transfer of isopentenyl residues [9, 13]. After the release of the diphosphate, the hemiterpenes are produced from DMAPP. Upon condensation of DMAPP and IPP via geranyl-PP synthase (dimethylallyl transferase), geranyl-PP forms. Monoterpenes are produced by geranyl-PP, which is the precursor of volatile oils. Likewise, farnesyl-PP is formed by condensation of geranyl-PP with IPP via farnesyl-PP synthase (geranyl transferase). Sesquiterpenes are produced by condensation of farnesyl-PP. In addition, two molecules of farnesyl-PP give rise to triterpenes [13, 14]. Squalene, a triterpenic compound, is the precursor of sterols that participate in membrane structure. Geranylgeranyl-PP is formed from head-to-tail condensation of farnesyl-PP and IPP by geranylgeranyl-PP synthase (farnesyl transferase). Geranylgeranyl-PP also causes the formation of diterpenes (**Figure 2**) [9].

Tetraterpenes (carotenoids) are formed by the condensation of two molecules of geranylgeranyl-PP [9, 15]. The first carotenoid formed by prephytoene diphosphate in this step is phytoene which is found in green plants in association with chlorophyll. This first stable carotenoid synthesis takes place via the phytoene synthase. The second product of carotenoid biosynthesis, phytofluene, forms as a result of desaturation (catalyzed by the enzyme phytoene desaturase), which leads to the formation of saturated double bonds. Phytofluene, after having a series of dehydrogenation reactions, forms a symmetric molecule, lycopene, containing 13 double bonds. The next step, α-carotene, β-carotene, and γ-carotene are produced from lycopene via ring formation in the end groups by lycopene cyclase. α- and β-carotenes are the xanthophylls as a result of hydroxylation by β-carotene hydroxylase (**Figure 2**) [16, 17].

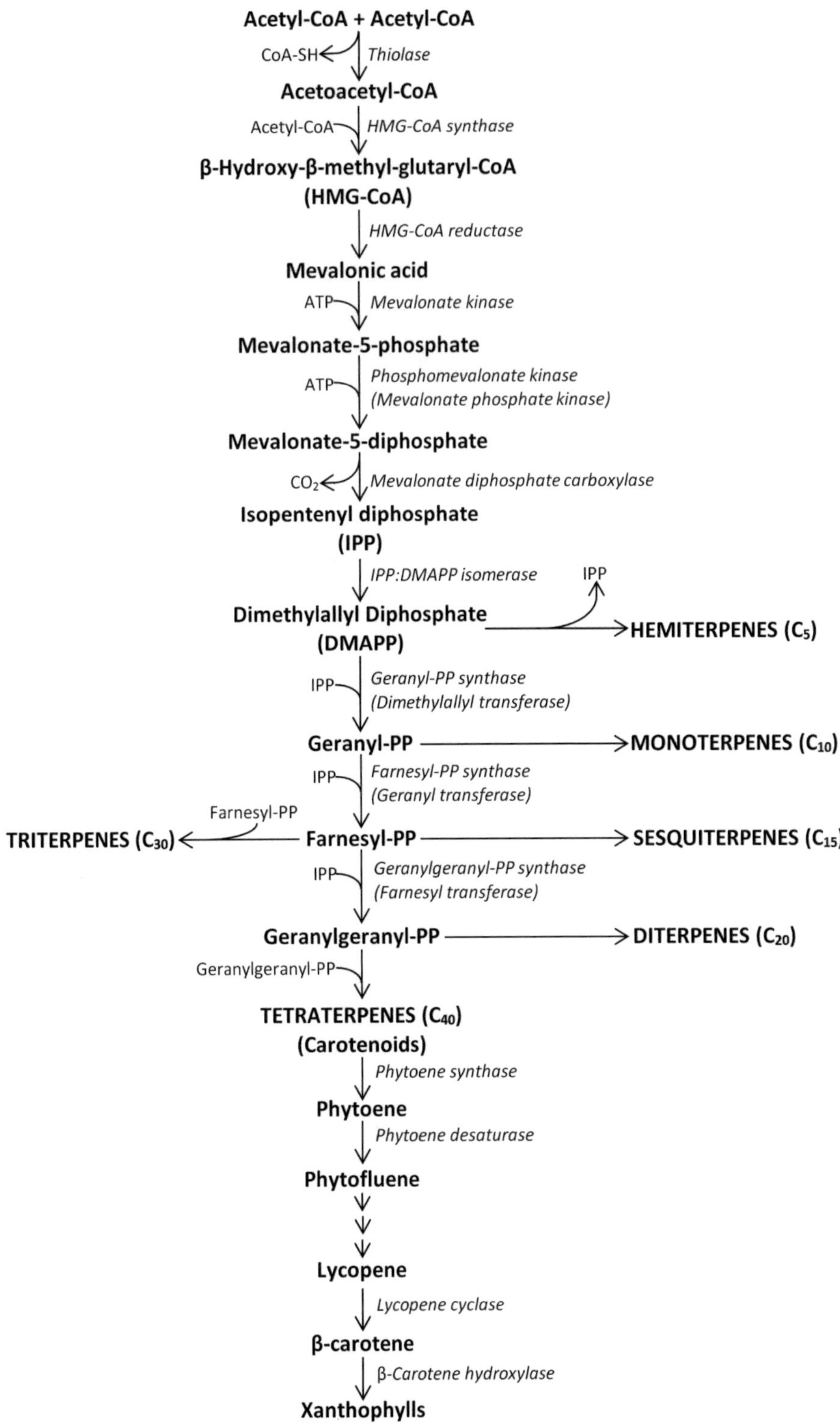

Figure 2. Biosynthesis of terpenes and carotenoids.

http://dx.doi.org/10.5772/67592

2.3. Carotenoids

More than 700 natural carotenoids have been identified, and new carotenoids have been added to this number every year [18–20]. Carotenoids are found in red-, yellow-, and orange-colored fruits and vegetables as well as in all green leafy vegetables [10].

Carotenoids are found in plant tissues as free forms (crystalline or amorphous) are dissolved form in an oily solvent. They may also be esterified with fatty acids or complexed with sugars and proteins [17]. The conjugate double-bond structure found in carotenoids also determines biological functions, such as absorption of light during photosynthesis, energy transfer, and protection from harmful effects of light on the cells during the photosynthesis. The presence of carotenoids also determines the characteristic color of these compounds [21].

Carotenoids are present in large quantities in the Leydig cells, which produce steroid hormones, and in the outer layer of the adrenal gland [22]. Carotenoids are also found in fresh broccoli, vegetables, butter, egg yolks, and animal-based foods [23].

2.4. Classification of carotenoids according to their structures

According to structures of carotenoids, two classes are distinguished as hydrocarbon carotenoids and xanthophylls (**Figure 3**) [24]. The apolar characteristics of hydrocarbon carotenoids are also called carotenes which include α-carotene, β-carotene, γ-carotene, lycopene, phytoene, and phytofluene. Xanthophylls are more polar and contain oxygen in the form of methoxy, hydroxy, keto, carboxy, and epoxy positions. Examples of xanthophylls include lutein, zeaxanthin, β-cryptoxanthin, astaxanthin, and fucoxanthin [25, 26]. Carotenoids are also grouped as carotenoids with or without ring groups at the ends of the chain [26].

2.4.1. Hydrocarbon carotenoids

Carotenoids with hydroaromatic rings are called carotenes [23]. These rings are located at the both ends of the four isoprene molecules. Thus, there are two hydroaromatic rings in each carotene molecule. These hydrocarbons are called ionospheric rings and have three ionone rings. These are α-, β-, and pseudo-ionone rings. α- and β-ionone rings are closed rings and include one double bond. The position of the double bond α is different in the β-ionone rings and pseudo rings. The pseudo-ionone ring includes two double bonds [3].

Carotene is a biochemically synthesized terpene from eight isoprene units [4, 8]. There are three main types, α-carotene, β-carotene, and γ-carotene. The carotenoids are precursors of vitamin A and converted into vitamin A in the body. From α-carotene and γ-carotene, one molecule of vitamin A while from β-carotene two molecules of vitamin A are synthesized [27–29].

α-Carotene molecule has β-ionone ring at one end and α-ionone ring at the other end. There are four molecules of polymerized isoprene molecules between them. The structure of vitamin A is like half of an α-carotene. Vitamin A has a β-ionone ring at one end and two isoprene molecules attached to it [27]. The second most common form of carotene, α-carotene, is found mostly in carrots, sweet potatoes, squash, tomatoes, red peppers [30], and dark green vegetables [26].

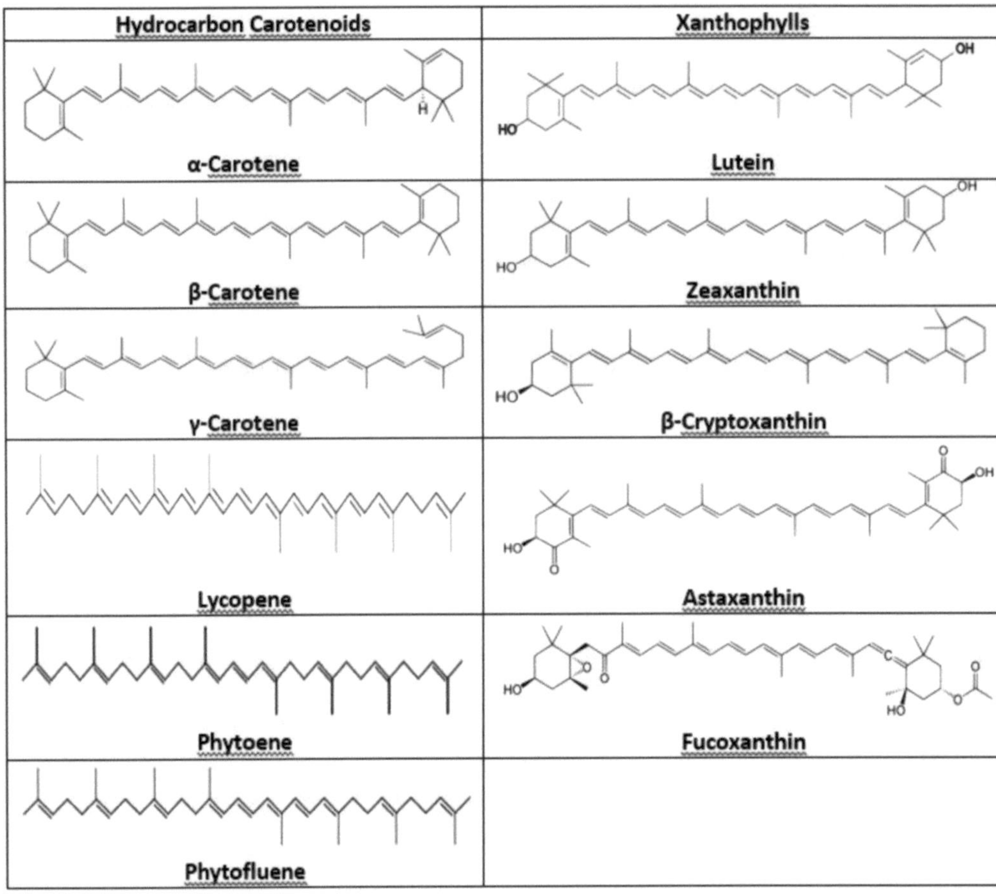

Figure 3. Classification of carotenoids according to their structures.

β-Carotene is a fat-soluble provitamin. Its active form is vitamin A [31]. The difference of β from *α* is that it carries the β-ionone ring at both ends. When β-carotene is divided into two molecules, vitamin A is synthesized [29]. It is found in fruits, cereals, vegetables (carrots, green plants, pumpkin, spinach), and oils [32].

γ-Carotene includes β-ionone ring at one end and the pseudo-ionone ring at the other end, and when the molecule is split, one molecule of vitamin A is synthesized [3].

Lycopene is an aliphatic carotenoid [33]. Lycopene is found among tomatoes, watermelons, pink grapefruit, and rosehip. Since lycopene dissolves in the oil, the presence of oils greatly increases its absorption by the digestive system [34].

Phytoene is a 40-carbon intermediate in the biosynthesis of phytoene carotenoids. Phytoene is a symmetric molecule containing three conjugated double bonds [35].

Phytofluene is an orange-colored carotenoid pigment found naturally in tomatoes and other vegetables [36].

2.4.2. *Xanthophylls*

Xanthophylls contain oxygen atoms and are yellow pigments commonly found in nature [37].

Lutein is a dihydroxy-carotene formed from carotenoids with an alcohol group containing hydroaromatic α structure [38]. Its both ionone rings carry hydroxyl groups [39]. Lutein is a substance that gives color to chicken fat, egg yolk, and chicken feathers [40]. That yellow-colored lutein found in plants is an organic colorant on leaves of green vegetables, such as spinach and black pepper. It is generally found in covalent interactions within fatty acids [20].

The chemical formulas of lutein and zeaxanthin are the same; in other words, they are isomers, but they are not stereoisomers. The difference between the two is the position of a double bond in the end ring [17, 39].

Zeaxanthin is one of the most common carotenoid alcohols found in the nature. It is a pigment that gives its color to maize, saffron, and many other plants. When the zeaxanthin breaks down, picrocrocin, which is responsible for the taste and aroma of the saffron, forms [41].

β-Cryptoxanthin carotenoid has an alcohol group having a hydroaromatic structure including an OH group in one of its ionic rings. Since the other ring is the β-ionone ring, one molecule of vitamin A can be synthesized from it [27].

β-Cryptoxanthin is a natural carotenoid pigment. β-Cryptoxanthin is found in fruits and vegetables, such as mandarin, red pepper, and zucchini, and has important functions for human health. β-Cryptoxanthin is closely related to β-carotene. Only one OH group was added to β-cryptoxanthin. Although β-carotene is present in large quantities in a large number of fruits and vegetables, β-cryptoxanthin is present in small number of food sources but at high concentrations [42].

Astaxanthin is a keto-carotenoid and is a zeaxanthin metabolite, containing both hydroxyl and ketone functional groups [17, 43]. Astaxanthin is found in microalgae, yeast, salmon, trout, shrimp, shellfish, and some birds' feathers [44].

Fucoxantin, found in the chloroplasts of algae and mosses, gives them their brown or olive green color [45].

3. Antioxidant functions of carotenoids at molecular level for health and toxicity

The interest in carotenoids found in plants over the last years is not only due to their A provitamin activity but also due to their reduction of oxidative stress in the organism by capturing oxygen radicals, that is, their antioxidant effects [46]. Free oxygen radicals play an important role in the stress-related tissue damage and pathogenesis of inflammation. The imbalance between protective and damaging mechanisms results in acute inflammation accompanied by neutrophil infiltration [47, 48]. Superoxide radicals formed by neutrophils react with lipids and cause lipid peroxidation [49–51]. Carotenoids can inhibit active radicals by transferring electrons, giving hydrogen atoms to radicals or attaching to radicals [16].

The number of conjugate double bonds in their structure is closely related to the superoxide inhibitory effect of carotenoids [52, 53]. Carotenoids could remove singlet oxygen and peroxyl radicals from reaction medium and also prevent their formation [53–55]. It is stated that carotenoids can inhibit cell renewal and transformation and regulate gene expression that plays a role in the formation of certain types of cancer. On the other hand, in some studies it has been shown that carotenoids could stimulate cancer in some cases. For example, it has been reported that in smokers, synthetic β-carotene does not create protective activity against lung cancer and cardiovascular diseases and even it fastens the progression of an aforementioned diseases [56–58].

The activity of carotenoids as antioxidants also depends on their interaction with other antioxidants, such as vitamins E and C [58]. In addition, some carotenoids and their metabolites activate the nuclear factor-erythroid 2-related factor-2 (Nrf2) transcription factor, which triggers antioxidant gene expression in certain cells and tissues [58, 59]. Thus, a number of chronic diseases characterized by oxidative stress, inflammation, and impaired mitochondrial function have been reported to reduce Nrf2 expression in some animal models. Elevation of Nrf2 has been shown to be effective in the prevention and treatment of many chronic inflammatory diseases, including various cardiovascular, renal, or pulmonary diseases; toxic liver damage; metabolic syndrome; sepsis; autoimmune disorders; inflammatory bowel disease; and HIV infection [58]. Prevention of low-density lipoproteins (LDL) oxidation by carotenoids has been suggested to be the basis of carotenoids' protective activity against coronary vascular disease [26].

The signaling pathways and molecules influenced by carotenoids to prevent various diseases, such as cancer and cardiovascular diseases, involve growth factor signaling members, cell cycle-associated proteins, differentiation-related proteins, retinoid-like receptors, antioxidant response element, nuclear receptors, AP-1 transcriptional complex, the Wnt/β-catenin pathway, angiogenic proteins, and inflammatory cytokines. During the treatment of cardiovascular and eye diseases and cancer, the dose and the exposure time of β-carotene, lycopene, lutein, and zeaxanthin have been reported to be crucial [37].

α-Carotene intake might decrease the development of non-Hodgkin lymphoma [60]. α-Carotene concentration in the blood may be associated with the development of various cancers [30].

β-Carotene, which is accepted as a provitamin of vitamin A (retinol) which is required for the fulfillment of visual functions [61], effects the oxidative damage that formed on cellular lipids, proteins, and DNA as a result of sunlight and causes the formation of erythema, premature aging of the skin, development of photodermatitis, and skin cancer. β-Carotene protects the skin from harmful effects of UV light by its prevention of reactive oxygen species formation and anti-inflammatory properties [62, 63].

Studies have shown that β-carotene's provitaminase activity or antioxidant properties prevent diseases, such as arthrosclerosis, cataracts, multiple sclerosis, and some types of cancers [64]. Moreover in recent studies, β-carotene has been proposed to decrease cell proliferation and induce apoptosis of various cancer cell lines by inhibiting calcium-/calmodulin-dependent protein kinase IV [65]. In another study, β-carotene has been reported to have anticancer stem cell actions on neuroblastoma, and this anticancer action is enhanced by retinoic acid receptor β [66].

Lycopene, synthesized by many plants and microorganisms, is an antioxidant that cannot be synthesized by animals and humans [67]. Conjugated dienes are active in creating antioxidant activity [46], and lycopene is reported to have a higher antioxidant capacity than other carotenoids, and in particular among carotenoids, lycopene inhibits the risk of prostate cancer [52, 68].

The use of lycopene can reduce the risk of cardiovascular disease, diabetes, osteoporosis, and prostate, esophagus, colorectal, and mouth cancer risk. Recent studies indicate that lycopene intake has protective functions against cardiovascular diseases by lowering high-density lipoprotein (HDL)-associated inflammation [69]. It was proposed that there is an inverse association between the occurrence of pancreatic cancer and dietary lycopene intake together with vitamin A and β-carotene [70]. β-Carotene 9′,10′-oxygenase which is a key enzyme for the metabolism of lycopene has been proposed to have an important roles to prevent prostate cancer progression by inhibiting NF-κB signaling [71].

In neurodegenerative diseases, lycopene has been reported to increase permeability of blood-brain barrier, and a significant reduction in lycopene levels in diseases, such as Parkinson's disease and vascular dementia, has been observed. It has also been suggested that lycopene bestow protection against amyotrophic lateral sclerosis (ALS) impairment in humans [67].

The phytoene and phytofluene found in the diet accumulate in the human skin [72]. The dehydration of these carotenoids has a protective effect on the skin due to its UV absorber, antioxidant, and anti-inflammatory properties [73].

Zeaxanthin is one of two carotenoids found in the retina. Zeaxanthin mostly found at the center of the macula and lutein mostly found at the peripheral retina [39, 74]. Lutein and zeaxanthin are responsible for the formation of yellow pigment in the retina. Yellow pigments play an active role in protecting the eye from light and can prevent retinal damage [56].

β-Cryptoxanthin acts as a chemopreventive agent against lung cancer. β-Cryptoxanthin has been reported to decrease lung cancer through downregulating neuronal nicotinic acetylcholine receptor α7/PI3K signaling pathway [75]. In addition to lung cancer, β-cryptoxanthin enhances the action of a chemotherapeutic agent, oxaliplatin, to treat colon cancer [76]. Moreover, high lycopene and β-cryptoxanthin including diet might protect against aggressive prostate cancer [77].

Astaxanthin has an important role in the treatment and prevention of certain diseases by its antitumor properties and protection against free radicals, oxidation of basic polyunsaturated fatty acids, and UV light effect on cell membranes [20, 78]. For example, high concentrations of astaxanthin may suppress mammary carcinoma [79]. In addition, astaxanthin has beneficial health effects against the formation of prostate carcinogenesis and tumor progression by reactivating the expression of Nrf2 and Nrf2-target genes through epigenetic modification and chromatin remodeling [80]. Astaxanthin has been shown to have antitumorigenic and anti-inflammatory effects on human lung cancer cell lines by inhibiting ERK1/2 activity [81].

In vitro studies indicate that fucoxanthin stimulates apoptosis and decreases proliferation and migration in glioma cancer cell lines U87 and U251 through Akt/mTOR and p38 pathway inhibition [82]. Fucoxanthin intake may have health benefits for the treatment of Alzheimer's disease by inhibiting acetylcholinesterase and enhancing brain-derived neurotrophic factor expression [83].

Supplementation of carotenoids by humans may have some beneficial biological actions against the formation of numerous diseases, such as cancer, cardiovascular diseases, or their skin. In addition to nutrient supplements in humans, carotenoids have applications in animal feed. Optimal health promoting actions of carotenoids depend on their proper doses, lengths of treatment, and combinations of carotenoids to maximize their effects. In this book chapter, latest findings on the biochemical and antioxidant activities of main carotenoids and their possible mechanisms of action will be presented.

Author details

Oguz Merhan

Address all correspondence to: oguzmerhan@hotmail.com

Department of Biochemistry, Faculty of Veterinary, Kafkas University, Kars, Turkey

References

[1] Domonkos I., Kis M., Gombos Z., Ughy B. Carotenoids, versatile components of oxygenic photosynthesis. Progress in Lipid Research. 2013;**52**:539–561. DOI: 10.1016/j.plipres. 2013.07.001

[2] Pattanaik B., Lindberg P. Terpenoids and their biosynthesis in cyanobacteria. Life. 2015;**5**:269–293. DOI: 10.3390/life5010269

[3] Ahrazem O., Gomez-Gomez L., Rodrigo MJ., Avalos J., Limon MC. Carotenoid cleavage oxygenases from microbes and photosynthetic organisms: features and functions. International Journal of Molecular Sciences. 2016;**17**:pii: E1781. DOI: 10.3390/ijms17111781

[4] Hirschberg J. Production of high-value compounds: carotenoids and vitamin E. Current Opinion in Biotechnology. 1999;**10**:186–191.

[5] Lichtenthaler HK. The 1-deoxy-d-xylulose-5-phosphate pathway of isoprenoid biosynthesis in plants. Annual Review of Plant Physiology and Plant Molecular Biology. 1999;**50**:47–65. DOI: 10.1146/annurev.arplant.50.1.47

[6] Varma N. Phytoconstituents and their mode of extractions: an overview. Research Journal of Chemical and Environmental Sciences. 2016;**4**:08–15.

[7] Dewick PM. The biosynthesis of C5-C25 terpenoid compounds. Natural Product Reports. 2002;**19**:181–222. DOI: 10.1039/b002685i

[8] Wagner KH., Elmadfa I. Biological relevance of terpenoids. overview focusing on mono-, di- and tetraterpenes. Annals of Nutrition and Metabolism. 2003;**47**:95–106. DOI: 10.1159/000070030

[9] Iriti M., Faoro F. Chemical diversity and defence metabolism: how plants cope with pathogens and ozone pollution. International Journal of Molecular Sciences. 2009;**10**:3371–3399. DOI: 10.3390/ijms10083371

[10] Gomez-Garcia Mdel R., Ochoa-Alejo N. Biochemistry and molecular biology of carotenoid biosynthesis in chili peppers (Capsicum spp.). International Journal of Molecular Sciences. 2013;**14**:19025–19053. DOI: 10.3390/ijms140919025.

[11] Oliver J., Palou A. Chromatographic determination of carotenoids in foods. Journal of Chromatography A. 2000;**881**:543–555.

[12] Lichtenthaler HK., Schwender J., Disch A., Rohmer M. Biosynthesis of isoprenoids in higher plant chloroplasts proceeds via a mevalonate-independent pathway. FEBS Letters. 1997;**400**:3271–3274.

[13] Tholl D. Biosynthesis and biological functions of terpenoids in plants. Advances in Biochemical Engineering/Biotechnology. 2015;**148**:63–106. DOI: 10.1007/10_2014_295

[14] Chappell J. The biochemistry and molecular biology of isoprenoid metabolism. Plant Physiology. 1995;**107**:1–6.

[15] Wendt KU., Schulz GE. Isoprenoid biosynthesis: manifold chemistry catalyzed by similar enzymes. Structure. 1998;**6**:127–133.

[16] Kopsell DA., Kopsell DE. Accumulation and bioavailability of dietary carotenoids in vegetable crops. Trends in Plant Science. 2006;**11**:499–507.

[17] Sajilata MG., Singhal RS., Kamat MY. The carotenoid pigment zeaxanthin—a review. Comprehensive Reviews in Food Science and Food Safety. 2008;**7**:29–49.

[18] Arvayo-Enriquez H., Mondaca-Fernandez I., Gortarez-Moroyoqui P., Lopez-Cervantes J., Rodriguez-Ramirez R. Carotenoids extraction and quantification: a review. Analytical Methods. 2013;**5**:2916–2924. DOI: 10.1039/C3AY26295B

[19] Amorim-Carrilho KT., Cepeda A., Fente C., Regal P. Review of methods for analysis of carotenoids. Trends in Analytical Chemistry. 2014;**56**:49–73.

[20] Mezzomo N., Ferreira SRS. Carotenoids functionality, sources, and processing by supercritical technology: a review. Journal of Chemistry. 2016;Article ID 3164312, 16 pages. DOI: 10.1155/2016/3164312.

[21] Deming DM., Erdman JW. Mammalian carotenoid absorption and metabolism. Pure and Applied Chemistry. 1999;**71**:2213–2223.

[22] Stahl W., Schwarz W., Sundquist AR., Sies H. Cis-trans isomers of lycopene and beta-carotene in human serum and tissues. Archives of Biochemistry and Biophysics. 1992;**294**:173–177.

[23] Ahamad MN., Saleemullah M., Shah HU., Khalil IA., Saljoqi AUR. Determination of beta carotene content in fresh vegetables using high performance liquid chromatography. Sarhad Journal of Agriculture. 2007;**23**:767–770.

[24] Armstrong GA., Hearst JE. Genetics and molecular biology of carotenoid pigment biosynthesis. FASEB Journal. 1996;**10**:228–237.

[25] Jaswir I., Noviendri D., Hasrini RF., Octavianti F. Carotenoids: sources, medicinal properties and their application in food and nutraceutical industry. Journal of Medicinal Plants Research. 2011;**5**:7119–7131. DOI: 10.5897/JMPRx11.011

[26] Eldahshan OA., Singab ANB. Carotenoids. Journal of Pharmacognosy and Phytochemistry. 2013;**2**:225–234.

[27] Harjes CE., Rocheford TR., Bai L., Brutnell TP., Kandianis CB., Sowinski SG. Natural genetic variation in lycopene epsilon cyclase tapped for maize biofortification. Science. 2008;**319**:330–333. DOI: 10.1126/science.1150255

[28] Grune T., Lietz G., Palou A., Ross AC., Stahl W., Tang G. β-carotene is an important vitamin A source for humans. The Journal of Nutrition. 2010;**140**:2268S–2285S. DOI: 10.3945/jn.109.119024

[29] Shete V., Quadro L. Mammalian metabolism of β-carotene: gaps in knowledge. Nutrients. 2013;**5**:4849–4868. DOI: 10.3390/nu5124849

[30] D'Adamo CR., Dawson VJ., Ryan KA., Yerges-Armstrong LM., Semba RD., Steinle NI. The CAPN2/CAPN8 locus on chromosome 1q Is associated with variation in serum alpha-carotene concentrations. Journal of Nutrigenetics and Nutrigenomics. 2016;**9**:254–264. DOI: 10.1159/000452890

[31] Valko M., Leibfritz D., Moncol J., Cronin MT., Mazur M., Telser J. Free radicals and antioxidants in normal physiological functions and human disease. The International Journal of Biochemistry & Cell Biology. 2007;**39**:44–84.

[32] Liu RH. Health-promoting components of fruits and vegetables in the diet. Advances in Nutrition. 2013;**4**:384S–392S. DOI: 10.3945/an.112.003517.

[33] Kildahl-Andersen G., Bruas L., Lutnaes BF., Liaaen-Jensen S. Nucleophilic reactions of charge delocalised carotenoid mono- and dications. Organic & Biomolecular Chemistry. 2004;**2**:2496–2506. DOI: 10.1039/B406913G

[34] Story EN., Kopec RE., Schwartz SJ., Harris GK. An update on the health effects of tomato lycopene. Annual Review of Food Science and Technology. 2010;**1**:16 pages. DOI: 10.1146/annurev.food.102308.124120.

[35] Sieiro C., Poza M., de Miguel T., Villa TG. Genetic basis of microbial carotenogenesis. International Microbiology. 2003;**6**:11–16. DOI: 10.1007/s10123-003-0097-0

[36] Engelmann NJ., Clinton SK., Erdman Jr JW. Nutritional aspects of phytoene and phytofluene, carotenoid precursors to lycopene. Advances in Nutrition: An International Review Journal. 2011;**2**:51–61. DOI: 10.3945/an.110.000075.

[37] Milani A., Basirnejad M., Shahbazi S., Bolhassani A. Carotenoids: biochemistry, pharmacology and treatment. British Journal of Pharmacology. 2016. DOI: 10.1111/bph.13625

[38] Kim J., DellaPenna D. Defining the primary route for lutein synthesis in plants: the role of Arabidopsis carotenoid beta-ring hydroxylase CYP97A3. Proceedings of the National Academy of Sciences of the United States of America. 2006;**103**:3474–3479. DOI: 10.1073/pnas.0511207103

[39] Abdel-Aal el-SM., Akhtar H., Zaheer K., Ali R. Dietary sources of lutein and zeaxanthin carotenoids and their role in eye health. Nutrients. 2013;**5**:1169–1185. DOI: 10.3390/nu5041169.

[40] McGraw KJ., Beebee MD., Hill GE., Parker RS. Lutein-based plumage coloration in songbirds is a consequence of selective pigment incorporation into feathers. Comparative Biochemistry and Physiology. Part B. 2003;**135**:689–696. DOI: 10.1016/S1096-4959(03)00164-7

[41] Paredi G., Raboni S., Marchesani F., Ordoudi SA., Tsimidou MZ., Mozzarelli A. Insight of saffron proteome by gel-electrophoresis. Molecules. 2016;**21**:167. DOI: 10.3390/molecules21020167

[42] Burri BJ., La Frano MR., Zhu C. Absorption, metabolism, and functions of β-cryptoxanthin. Nutrition Reviews. 2016;**74**:69–82. DOI: 10.1093/nutrit/nuv064

[43] Ambati RR., Phang SM., Ravi S., Aswathanarayana RG. Astaxanthin: sources, extraction, stability, biological activities and its commercial applications—a review. Marine Drugs. 2014;**12**:128–152. DOI: 10.3390/md12010128

[44] Dalei J., Sahoo D. Extraction and characterization of astaxanthin from the crustacean shell waste from shrimp processing industries. International Journal of Pharmaceutical Sciences and Research. 2015;**6**:2532–2537. DOI: 10.13040/IJPSR.0975-8232.6(6).2532-37

[45] Mikami K., Hosokawa M. Biosynthetic pathway and health benefits of fucoxanthin, an algae-specific xanthophyll in brown seaweeds. International Journal of Molecular Sciences. 2013;**14**:13763–13781. DOI: 10.3390/ijms140713763

[46] Rao AV., Agarwal S. Role of lycopene as antioxidant carotenoid in the prevention of chronic diseases: a review. Nutrition Research. 1999;**19**:305–323.

[47] Moldoveanu B., Otmishi P., Jani P., Walker J., Sarmiento X., Guardiola J. Inflammatory mechanisms in the lung. Journal of Inflammation Research. 2009;**2**:1–11.

[48] Mittal M., Siddiqui MR., Tran K., Reddy SP., Malik AB. Reactive oxygen species in inflammation and tissue injury. Antioxidants & Redox Signaling. 2014;**20**:1126–1167. DOI: 10.1089/ars.2012.5149

[49] Valko M., Rhodes CJ., Moncol J., Izakovic M., Mazur M. Free radicals, metals and antioxidants in oxidative stress-induced cancer. Chemico-Biological Interactions. 2006;**160**:1–40. DOI: 10.1016/j.cbi.2005.12.009

[50] Pham-Huy LA., He H., Pham-Huy C. Free radicals, antioxidants in disease and health. International Journal of Biomedical Science. 2008;**4**:89–96.

[51] Nimse SB., Pal D. Free radicals, natural antioxidants, and their reaction mechanisms. RSC Advances. 2015;**5**:27986. DOI: 10.1039/c4ra13315c

[52] Stahl W., Sies H. Bioactivity and protective effects of natural carotenoids. Biochimica et Biophysica Acta. 2005;**1740**:101–107. DOI: 10.1016/j.bbadis.2004.12.006

[53] Merhan O., Ozcan A., Atakisi E., Ogun M., Kukurt A. The effect of β-carotene on acute phase response in diethylnitrosamine given rabbits. Kafkas Univ Vet Fak Derg. 2016;**22**:533–537. DOI: 10.9775/kvfd.2016.14995

[54] Chaudiere J., Ferrari-Iliou R. Intracellular antioxidants: from chemical to biochemical mechanisms. Food and Chemical Toxicology. 1999;**37**:949–962. DOI: 10.1016/S0278-6915(99)00090-3

[55] Young IS., Woodside JV. Antioxidants in health and disease. Journal of Clinical Pathology. 2001;**54**:176–186. DOI: 10.1136/jcp.54.3.176

[56] Tapiero H., Townsend DM., Tew KD. The role of carotenoids in the prevention of human pathologies. Biomedicine & Pharmacotherapy. 2004;**58**:100–110. DOI: 10.1016/j.biopha.2003.12.006

[57] Touvier M., Kesse E., Clavel-Chapelon F., Boutron-Ruault MC. Dual association of beta-carotene with the risk of tobacco-related cancers in a cohort of French women. Journal of the National Cancer Institute. 2005;**97**:1338–1344. DOI: 10.1093/jnci/dji276.

[58] Gammone MA., Riccioni G., D'Orazio N. Marine carotenoids against oxidative stress: effects on human health. Marines Drugs. 2015;**13**:6226–6246. DOI: 10.3390/md13106226.

[59] Pall ML., Levine S. Nrf2, a master regulator of detoxification and also antioxidant, anti-inflammatory and other cytoprotective mechanisms, is raised by health promoting factors. Acta Physiologica Sinica. 2015;**67**:1–18. DOI: 10.13294/j.aps.2015.0001

[60] Chen F., Hu J., Liu P., Li J., Wei Z., Liu P. Carotenoid intake and risk of non-Hodgkin lymphoma: a systematic review and dose-response meta-analysis of observational studies. Annals of Hematology. 2016. DOI: 10.1007/s00277-016-2898-1

[61] von Lintig J. Provitamin A metabolism and functions in mammalian biology. The American Journal of Clinical Nutrition. 2012;**96(suppl)**:1234S–1244S. DOI: 10.3945/ajcn.112.034629

[62] Stahl W., Sies H. Carotenoids and flavonoids contribute to nutritional protection against skin damage from sunlight. Molecular Biotechnology. 2007;**37**:26–30. DOI: 10.1007/s12033-007-0051-z

[63] Cazzonelli CI. Carotenoids in nature: insights from plants and beyond. Functional Plant Biology. 2011;**38**:833–847.

[64] Terao J. Antioxidant activity of beta-carotene-related carotenoids in solution. Lipids. 1989;**24**:659–661.

[65] Naz H., Khan P., Tarique M., Rahman S., Meena A., Ahamad S. Binding studies and biological evaluation of β-carotene as a potential inhibitor of human calcium/calmodulin-dependent protein kinase IV. International Journal of Biological Macromolecules. 2016;**96**:161–170. DOI: 10.1016/j.ijbiomac.2016.12.024

[66] Kim YS., Kim E., Park YJ., Kim Y. Retinoic acid receptor β enhanced the anti-cancer stem cells effect of β-carotene by down-regulating expression of delta-like 1 homologue in human neuroblastoma cells. Biochemical and Biophysical Research Communications. 2016;**480**:254–260. DOI: 10.1016/j.bbrc.2016.10.041

[67] Rao AV., Rao LG. Carotenoids and human health. Pharmacological Research. 2007;**55**:207–216. DOI: 10.1016/j.phrs.2007.01.012

[68] Giovannucci E., Ascherio A., Rimm EB., Stampfer MJ., Colditz GA., Willett WC. Intake of carotenoids and retinol in relation to risk of prostate cancer. Journal of the National Cancer Institute. 1995;**87**:1767–1776.

[69] Thies F., Mills LM., Moir S., Masson LF. Cardiovascular benefits of lycopene: fantasy or reality? The Proceedings of the Nutrition Society. 2016:1–8. DOI: 10.1017/S0029665116000744

[70] Huang X., Gao Y., Zhi X., Ta N., Jiang H., Zheng J. Association between vitamin A, retinol and carotenoid intake and pancreatic cancer risk: evidence from epidemiologic studies. Scientific Reports. 2016;**6**:38936. DOI: 10.1038/srep38936

[71] Gong X., Marisiddaiah R., Zaripheh S., Wiener D., Rubin LP. Mitochondrial β-carotene 9',10' oxygenase modulates prostate cancer growth via NF-κB inhibition: a lycopene-independent function. Molecular Cancer Research. 2016;**14**:966–975. DOI: 10.1158/1541-7786.MCR-16-0075

[72] Hsu BY., Pu YS., Inbaraj BS., Chen BH. An improved high performance liquid chromatography-diode array detection-mass spectrometry method for determination of carotenoids and their precursors phytoene and phytofluene in human serum. Journal of Chromatography B. 2012;**899**:36–45. DOI: 10.1016/j.jchromb.2012.04.034

[73] Aust O., Stahl W., Sies H., Tronnier H., Heinrich U. Supplementation with tomato-based products increases lycopene, phytofluene, and phytoene levels in human serum and protects against UV-light-induced erythema. International Journal for Vitamin and Nutrition Research. 2005;**75**:54–60. DOI: 10.1024/0300-9831.75.1.54

[74] Koo E., Neuringer M., SanGiovanni JP. Macular xanthophylls, lipoprotein-related genes, and age-related macular degeneration. The American Journal of Clinical Nutrition. 2014;**100**:336S–346S. DOI: 10.3945/ajcn.113.071563

[75] Iskandar AR., Miao B., Li X., Hu KQ., Liu C., Wang XD. β-cryptoxanthin reduced lung tumor multiplicity and inhibited lung cancer cell motility by downregulating nicotinic acetylcholine receptor $\alpha 7$ signaling. Cancer Prevention Research. 2016;**9**:875–886. DOI: 10.1158/1940-6207.CAPR-16-0161

[76] San Millan C., Soldevilla B., Martin P., Gil-Calderon B., Compte M., Perez-Sacristan B. β-Cryptoxanthin synergistically enhances the antitumoral activity of oxaliplatin through ΔNP73 negative regulation in colon cancer. Clinical Cancer Research. 2015;**21**:4398–4409. DOI: 10.1158/1078-0432.CCR-14-2027

[77] Antwi SO., Steck SE., Su LJ., Hebert JR., Zhang H., Craft NE. Carotenoid intake and adipose tissue carotenoid levels in relation to prostate cancer aggressiveness among African-American and European-American men in the North Carolina-Louisiana prostate cancer project (PCaP). Prostate. 2016;**76**:1053–1066. DOI: 10.1002/pros.23189

[78] Hu ZC., Zheng YG., Wang Z., Shen YC. pH control strategy in astaxanthin fermentation bioprocess by *Xanthophyllomyces dendrorhous*. Enzyme and Microbial Technology. 2006;**39**:586–590. DOI: 10.1016/j.enzmictec.2005.11.017

[79] Yuri T., Yoshizawa K., Emoto Y., Kinoshita Y., Yuki M., Tsubura A. Effects of dietary xanthophylls, canthaxanthin and astaxanthin on N-methyl-N-nitrosourea-induced rat mammary carcinogenesis. In Vivo. 2016;**30**:795–800.

[80] Yang Y., Fuentes F., Shu L., Wang C., Pung D., Li W. Epigenetic CpG methylation of the promoter and reactivation of the expression of GSTP1 by astaxanthin in human prostate LNCaP Cells. The AAPS Journal. 2016. DOI: 10.1208/s12248-016-0016-x

[81] Liao KS., Wei CL., Chen JC., Zheng HY., Chen WC., Wu CH. Astaxanthin enhances pemetrexed-induced cytotoxicity by downregulation of thymidylate synthase expression in human lung cancer cells. Regulatory Toxicology and Pharmacology. 2016;**81**:353–361. DOI: 10.1016/j.yrtph.2016.09.031

[82] Liu Y., Zheng J., Zhang Y., Wang Z., Yang Y., Bai M. Fucoxanthin activates apoptosis via inhibition of PI3K/Akt/mTOR pathway and suppresses invasion and migration by restriction of p38-MMP-2/9 pathway in human glioblastoma cells. Neurochemical Research. 2016;**41**:2728–2751. DOI: 10.1007/s11064-016-1989-7

[83] Lin J., Huang L., Yu J., Xiang S., Wang J., Zhang J. Fucoxanthin, a marine carotenoid, reverses scopolamine-induced cognitive impairments in mice and inhibits acetylcholinesterase in vitro. Marine Drugs. 2016;**14**:67. DOI: 10.3390/md14040067

Chapter 5

β-Carotene and Free Radical Reactions with Nitrogen Oxides

Sara N. Mendiara and Luis J. Perissinotti

Additional information is available at the end of the chapter

http://dx.doi.org/10.5772/67683

Abstract

The following presentation is based on experimental work we have already developed and published. We investigated the nitrogen oxides in different solvents and analyzed their reaction with β-carotene. The electron paramagnetic resonance spectroscopy (EPR) and ultraviolet and visible (UV-vis) spectroscopy were applied to investigate the reaction of β-carotene with nitrogen dioxide and nitric oxide in both pure dioxane and dioxane/water solvent. Free radicals were detected and evaluated with the EPR technique, which is highly selective and sensitive. A reaction mechanism was proposed on the basis of the experimental kinetic and EPR results. The validity of the mechanism was checked by applying simulation set up conditions that reproduced the results achieved. The radical intermediates proposed in the reaction: the β-carotene neutral radicals and the cyclic nitroxide neutral radicals were theoretically studied. For that purpose, the density functional theory (DFT) level was applied, selecting the most suitable method, the unrestricted Becke-style 3-parameter with the Lee-Yang-Parr correlation functional (UB3LYP) and the 6-31G(d) basis sets (d orbital functions). We developed an appropriate discussion on the importance of carotenoids compounds and their reactions in biological media. Also, we evaluated the role and the possible reactions of nitroxide intermediates.

Keywords: β-carotene, EPR, neutral radicals, nitrogen dioxide, nitroxides

1. Introduction

The following presentation is based on experimental work we have already developed and published. We investigated the behavior of nitrogen oxides in different solvents and analyzed the reaction of nitrogen oxides with β-carotene.

The reaction of free radicals with carotenoids and the properties of the carotenoid-free radicals formed are of widespread interest because of their potential role in biological systems. We have carried out our work only with β-carotene, an unsaturated and extensively conjugated hydrocarbon. In vitro studies had shown the potential of carotenoids to act as free-radical scavengers. Nitrogen oxides, such as nitric oxide (NO) and the nitrogen dioxide (NO_2), constitute a source of free radicals; they are species that have an unpaired electron. It is expected that nitrogen oxides may react in some important way with carotenoids and their radical intermediates. It is important to note that in our preliminary investigations, when mixtures of NO and NO_2 were added to certain organic purified compounds, compounds of the type of the nitroxides were detected using the technique of EPR [1].

Carotenoids are a family of pigmented compounds that are synthesized by plants and microorganisms but not animals. However, humans and primates accumulate them in several tissues. Carotenoids are absorbed in the intestinal mucosa like other lipophilic components. β-carotene molecule is geared in the lipophilic membranes interacting, through van der Waals forces, with the hydrocarbon chains of the lipids [2–5]. Carotenoids can be traced in the cellular cytoplasm where they can interact with different components, including the nitrogen oxides. It is also important to take into account the metabolism of carotenoids; β-carotene is particularly hydrophobic and it is reasonable to hypothesize that it would need to be transformed to more polar compounds in order to be excreted via urine [6]. As well, lutein and β-carotene were measured in human brain tissue and related to better cognition in octogenarians. The protective effect may not merely be an antioxidant effect given that α-tocopherol was less related to cognition than carotenoids [7]. The first report of the presence of β-carotene in the brain was in 1976, the patient was taking a high-dose β-carotene as a treatment and the carotenoid was measured within whole sections of the cerebrum [8].

We understood that the research might be quite profitable. The purpose of our work was to study and to enlighten the following problems:

(a) The behavior of nitrogen oxides: NO and NO_2 in some solvents (2001) [1].

(b) The interaction and reaction among nitrogen oxides and β-carotene (2009) [9].

(c) Theoretical evaluation of the intermediates proposed: the acyclic β-carotene neutral radicals, the acyclic nitrous β-carotene neutral radicals, and the cyclic nitroxide neutral radicals (2015) [10].

2. Development: (a), (b), and (c): conclusions and perspective

2.1. (a) The behavior of the nitrogen oxides NO and NO_2 in some solvents

Luckily, we had started the research developing a detailed study of solutions of nitrogen oxides in several solvents [1]. Those solutions were monitored with the following spectroscopic techniques: ultraviolet and visible (UV-vis) and electron paramagnetic resonance (EPR). The research allowed us to know what solvents can be used because they do not react with nitrogen oxides. We learned how to develop a careful control of the solvents. The knowledge acquired in this first work allowed us to follow, understand, and manage to clear up the results later finally achieved from the β-carotene assays. Therefore, the solvent dioxane and other later used in our research were adequately tested. It was also verified that they did not react with the nitrogen oxides, NO_2 and NO, even after prolonged times of observation in EPR (unlike what happened with hexane) [9].

NO_2 solutions were very carefully followed. In this chapter, we show only highlights from the developed research. Our first manuscript on nitrogen oxides, carried out in 2001, showed experimental and technical research in detail. **Figure 1** displays a set of EPR spectral records of the radical NO_2 at environmental temperature. Solutions of NO_2 were prepared in different

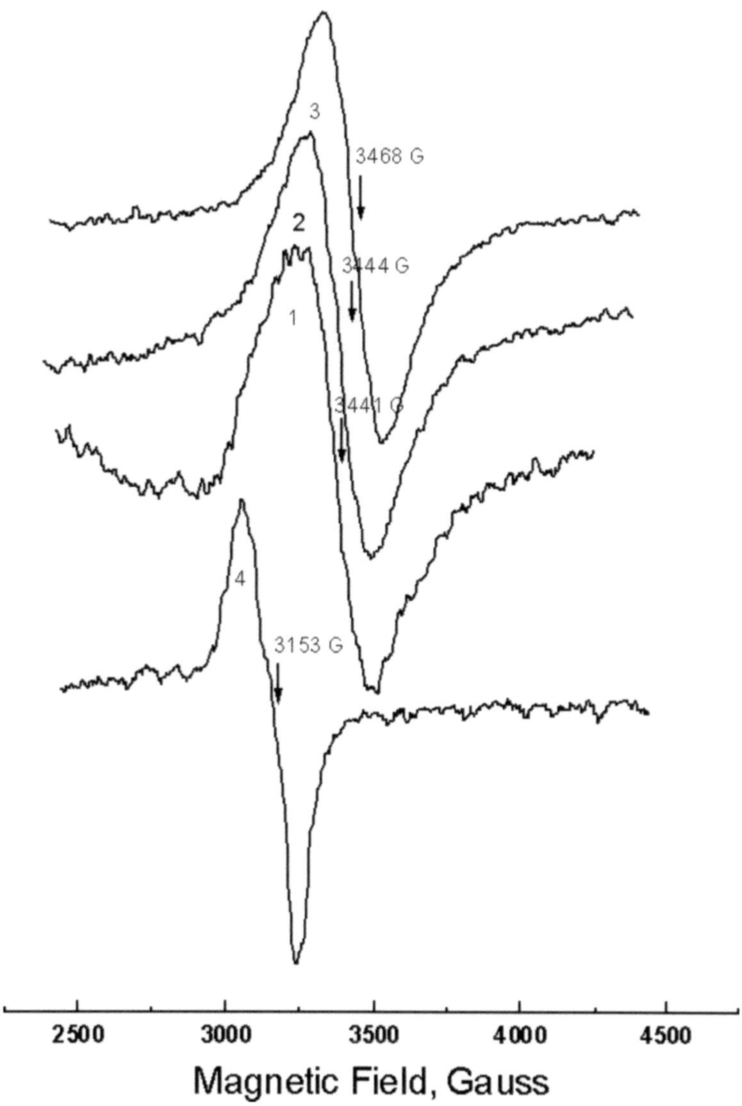

Figure 1. EPR spectra of NO_2, at 290 K, in: (1) gas-phase; (2) hexane; (3) carbon tetrachloride; and (4) acetone. General instrumental settings: microwave power, 18,000 mW; attenuation, 9 dB; modulator frequency, 100 kHz; modulation amplitude, 2.5 G; time constant or response time, 0.2 s; scan rate, 200 s. The recipient was previously flushed with argon.

media: hexane, benzene, carbon tetrachloride, acetone, and water. You can see that each spectrum do not differ considerably from the others. The signal is not resolved, due to the movement of the free radical and to the collisions that usually take place at the working temperature. Noticed that our research was always carried out under argon atmosphere [1, 9].

In **Table 1**, we can see the values of g, the spectroscopic factor and of ΔB, the total spectrum line width. Those data were extracted or calculated from **Figure 1**. We can also observe that the values in non-polar media are near the result measured in the gaseous phase. The general instrumental settings were microwave power, 18.000 mW; attenuation, 9 dB; modulator frequency, 100 kHz; modulation amplitude, 2.5 G; time constant or response time, 0.2 s; scan rate, 200 s.

We also prepared NO solutions. The preparation and handling of solutions of NO were performed under strict inert atmosphere. Even though precautions were employed to exclude oxygen from the system, a fingerprint UV-vis spectrum assigned to the presence of $NO_2/N_2O_4/HNO_2$ as impurity was sometimes observed. UV-vis spectra of solutions of NO/NO_2 in different solvents were analyzed. Those Spectra shown splits in some solvents. Some authors had interpreted them as the spectrum of NO [11]. However, it was confirmed that those UV-vis spectra were obtained in polar solvents and could be attributed to the formation of HNO_2 [1, 12].

Usually, one has to control nitrogen dioxide samples, recording the UV-vis spectrum and the corresponding EPR spectrum. The EPR spectrum of NO_2 is looked for in the following field interval: 1000–2000 Gauss, as you can observe in **Figure 1**. If we want to search for nitroxides, we must work in the following scan range: 50–200 Gauss. The EPR spectrum of NO was not detected in our field interval work.

When NO was dissolved in hexane, nitroxide intermediates were detected after 20 minutes or more, the radical was persistent. Probably no reaction took place when pure NO was present, but a trace amount of NO_2 initiated the reaction [13, 14].

In 2003, we carried out the measurement of the N_2O_4 dissociation constant ($N_2O_4/2\ NO_2$) in some solvents [15]. The N_2O_4 dissociation constant measured in hexane, carbon tetrachloride,

Medium	g [a]	ΔB, Gauss
Gas phase	1.999	1190 ± 58
Hexane	1.981	1194 ± 40
Benzene	1.985	563 ± 20
Carbon tetrachloride	1.978	897 ± 34
Acetone	2.185	499 ± 23
Water	2.176	502 ± 20

*Results obtained in the present work. Instrumental settings are shown in **Figure 1**.

[a]g values were determined with reference to a diphenylpicrylhydrazyl standard, g = 2.0036. The absolute errors are 0.002–0.004.

Table 1. EPR spectroscopic data of NO_2 radical at 290 K: spectroscopic factor, g, and total line width, ΔB*.

and chloroform compares approximately with the values calculated by some previous authors [16–18]. The techniques usually applied were colorimetric and spectrophotometric ones. As the absorption along the appropriate wavelength range was small, the errors in measurements at low concentrations were considerably large. The EPR technique has also quantification errors; however, the method is more reliable and the radical is directly detected. As far as we were aware, this was the first attempt to measure this equilibrium with the EPR technique.

2.2. (b) The interaction and reaction among nitrogen oxides and β-carotene: kinetic analysis and the corresponding modeling work

2.2.1. *UV-vis measurements*

β-Carotene, a non-polar molecule, is insoluble in polar solvents like water. We desired to investigate the reaction of β-carotene with the nitrogen oxides generated in situ from aqueous solutions of sodium nitrite and sulfuric acid. In order to achieve our purpose, we prepared solutions of β-carotene in pure and adequately checked dioxane solvent, under argon atmosphere and constant temperature. The dioxane-water mixtures remained homogeneous and so the UV-vis spectroscopy studies could be applied. Other aprotic solvents easily cause cloudiness when water is added, then UV-vis measurements cannot be carried out.

2.2.2. *EPR measurements*

EPR measurements were carried out with the β-carotene solution in pure dioxane plus an aliquot of NO and NO_2 in pure dioxane. The solutions in dioxane had low concentration of the persistent radical, besides the triplet-type signal was recorded during 5 days. In pure dioxane, we only developed a qualitative approach. On the other hand, we could design a quantitative study carried out with β-carotene solution in pure dioxane plus an aliquot of the dioxane/water solution with the nitrogen oxides formed in situ from sodium nitrite in an acid medium. Also the formation and decay of the persistent intermediates was monitored and shown in **Figure 2**. Dioxane solvent was tested, and no EPR signal was detected even after prolonged times of observation.

2.2.3. *Kinetic measurements*

Kinetic measurements were developed in dioxane/water solvent at 298 K. All the measurements were successfully simulated with the software for chemical kinetics, following the reaction path proposed in **Figure 3**. In **Table 2**, each reaction was described and the corresponding kinetic constant assigned. Unfortunately, in our 2009 manuscript, the sixth equation had one mistake, we must write P_1 instead of P_2 [9]. In the 2015 manuscript, the rate constants k_i (i = 1 to 7) were not written above the arrows [10].

In **Figure 3** and **Table 2**, we observed that whenever both NO and NO_2 were present, abstraction took place and nitroxides formed. Furthermore, it has already been shown that NO is less

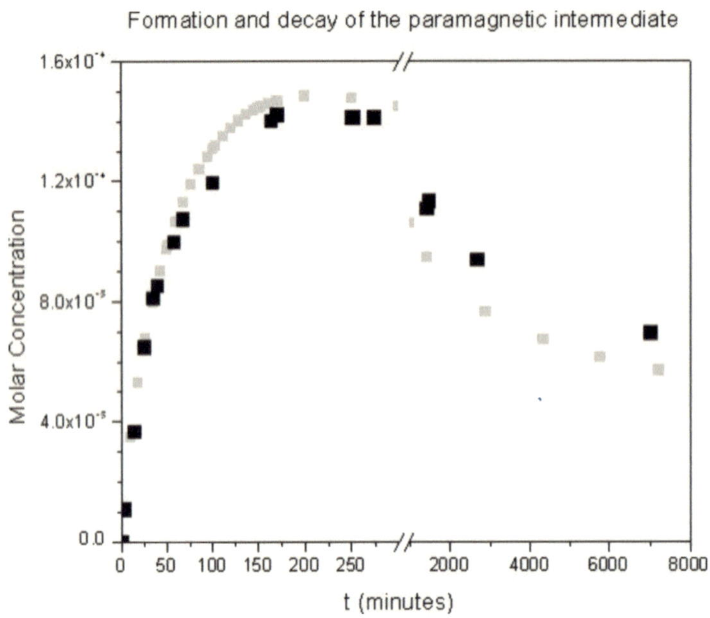

Figure 2. Formation and decay of a persistent type radical generated from a solution of β-carotene, 1×10^{-2}M, and nitrite anion, 1×10^{-1}M in dioxane:water ($x_{dioxane} = 0.65$) and in the presence of an acid medium, pH = 2 at 298 K. The gray curve and symbols represent the values obtained from the kinetic simulation process; the black square symbols represent our experimental values. After 20 days, we were still able to measure a radical concentration of 3.5×10^{-5} M, not shown in the graph. The reactions were carried out under argon atmosphere.

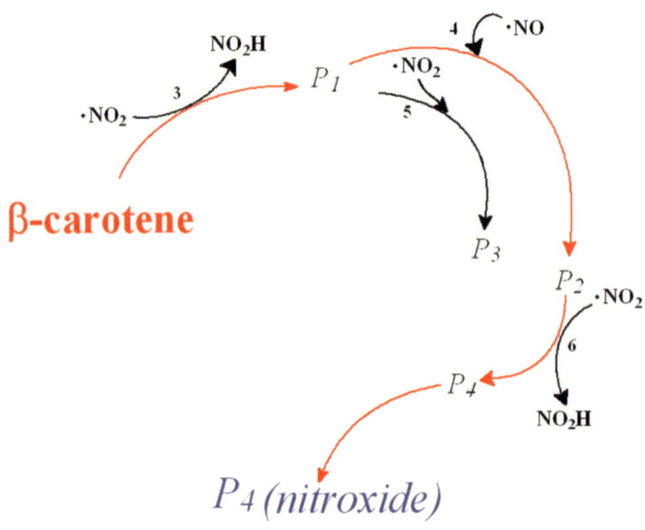

Figure 3. General reaction pathway proposed. The figure shows, in first place, the abstraction reactions of allylic hydrogens from β-carotene by nitrogen dioxide radicals and the formation of the products P_1. P_1 represents different possible β-carotene neutral radicals. P_2 represents the nitrous β-carotene formed with nitric oxide radicals. P_3 represents the products formed with nitrogen dioxide radicals, not studied in this work. P_4 represents the neutral radicals that are formed after the reaction of P_2 with nitrogen dioxide radicals, another abstraction reaction of allylic hydrogens has taken place. Finally, when P_4 generates an internal ring, a radical $P_{4(nitroxide)}$ comes up.

Kinetic analysis in acid medium[a]	
$2NO_2H \xrightarrow{k_1} NO + NO_2 + H_2O$	$k_1 = 1.76 \times 10\ M^{-1}s^{-1}$
$NO + NO_2 + H_2O \xrightarrow{k_{-1}} 2NO_2H$	$k_{-1} = 3.01 \times 10^6\ M^{-1}s^{-1}$
$2NO_2 + H_2O \xrightarrow{k_2} NO_2H + NO_3^- + H^+$	$k_2 = 1.54 \times 10^6\ M^{-1}s^{-1}$
$\beta-carotene + NO_2 \xrightarrow{k_3} P_1 + NO_2H$	$k_3 = 4.7 \times 10^5\ M^{-1}s^{-1}$
$P_1 + NO \xrightarrow{k_4} P_2$	$k_4 = 1.0 \times 10^9\ M^{-1}s^{-1}$
$P_1 + NO_2 \xrightarrow{k_5} P_3$	$k_5 = 1.0 \times 10^9\ M^{-1}s^{-1}$
$P_2 + NO_2 \xrightarrow{k_6} P_4 + NO_2H$	$k_6 = 7.6\ M^{-1}s^{-1}$
$P_4 + P_4(nitroxide) \xrightarrow{k_7} recombination; desproportionation$	$k_7 = 2.5 \times 10^{-1}\ M^{-1}s^{-1}$

[a]T = 298K. k_1, k_{-1}, and k_2 were already known. The rate constants k_4 and k_5 were considered diffusion controlled. The rate constants k_3, k_6, and k_7 were deduced from our modeling work. The kinetic simulation reproduced successfully the experimental kinetics results [9].

P_1 represents a possible allylic-type radical. P_2 represents the possible isomers molecules of nitrous carotene. P_4 represents the possible allylic-type radicals obtained from P_2. When a radical P_4 cycles internally a radical P_4*(nitroxide)* is formed. P_3 represents the products formed with nitrogen dioxide, not studied in this work.

Table 2. Reaction of β-carotene and nitrite in dioxane/water solution.

reactive and less efficient in the abstraction of a hydrogen atom than NO_2 [9]. We can compare the following bond dissociation energies (BDE) values at 298 K:

- BDE (H-NO) = 195.35 ± 0.25 kJ.mol^{-1}
- BDE (H-NO_2 or H-ONO) = 327.6 ± 2.1 kJ.mol^{-1}

We proposed the formation of β-carotene neutral radicals that in the presence of nitrogen oxides, originated persistent radical intermediates. The reactions were carried out under argon atmosphere because of the reactivity of triplet oxygen; otherwise oxygenated compounds would be formed with the carotenoid neutral radicals.

We followed and measured the difficult kinetic of the mechanism proposed. The kinetic of the reaction depended on β-carotene, nitrite, and acid concentrations. We experimentally verified that the β-carotene followed a first-order decay and we measured the corresponding pseudo-first-order-constant [9].

In **Table 2**, a set of reactions represents the mechanism of the reaction. It is assumed that the system always behaves as if it has reached the acid-base balance.

P_4 was considered as P_4*(nitroxide)*. We ran a non-commercial simulation program of chemical kinetics. The validity of the proposed mechanism was therefore tested by numerical integration.

The values of k_1, k_{-1}, *and* k_2 were extracted from the literature. The rate constant k_3 was adjusted following the UV-vis decay of β-carotene. The rate constants k_6 and k_7 were involved in the formation of the persistent radicals and were adjusted to the EPR results. The rate constants k_4 and k_5 were the recombination reactions between P_1 radicals and nitrogen dioxide and nitric oxide radicals. We calculated the following values: $k_4 = 1.4\times10^{10}M^{-1}s^{-1}$ and $k_5 = 1.1\times10^{10}M^{-1}s^{-1}$, by applying the equations of Smoluchoski, Stokes, and Einstein [19]. The calculations were developed taking into account the viscosity of the mixture dioxane-water, $\eta_{25°C} = 1.42 \times 10^{-3}$ Pa.s and the molecular size of the NO, NO_2 and β-carotene neutral radicals [20]. Actually, these constants were smaller because the nitrogen oxides reacted at some selected regions of the β-carotene neutral radicals. It was confirmed that changes of k_4 and k_5 in the order of 10^7 to 10^{11} had no significant effect upon the simulation results, which match very well with our experimental results. So the value of 1.0×10^9 $M^{-1}s^{-1}$ was selected and shown for k_4 and k_5 in **Table 2**. P_3 represents the products formed with nitrogen dioxide, not studied in this work.

We considered that the nitrous acid was proportional to the nitrite anion through the acid-base equilibrium [9]. The simulation results at 298 K reproduced quite well the experimental decay of β-carotene in the range near the half-life at pH = 5.3. The formation and the decay of the persistent radical intermediate at pH = 2.2 were also very well reproduced. For example, we showed *Case (a)* and *Case (b)* through which we tested the experiments carried out.

Case (a) Initial concentrations: β-carotene, 8.0×10^{-6}M; nitrite anion, 9.3×10^{-3}M; pH, 5.3

The simulation program showed that the system achieved approximately the following steady state order of concentrations: $[NO_2] \cong 10^{-8}$ to 10^{-9}M $[NO] \cong 10^{-6}$ to 10^{-7}M $[P_1] \cong 10^{-10}$M $[P_2] \cong 10^{-6}$M

The pH achieved in the simulation run remained almost constant: 5.30 ± 0.02.

The maximum radical concentration reached was: $[P_4 + P_{4(nitroxide)}] \cong 10^{-11}$M, in agreement with our EPR results.

The persistent radicals were not detected; evidently, our equipment had not enough sensibility.

Case (b) Initial concentrations: β-Carotene, 1.0×10^{-2}M; nitrite anion, 1.0×10^{-1}M; pH, 2.2.

The system achieved approximately the following steady state concentrations order: $[NO_2] \cong 10^{-7}$M $[NO] \cong 10^{-2}$M $[P_1] \cong 0$M $[P_2] \cong 10^{-3}$M.

The pH achieved in the simulation run remained almost constant: 2.20 ± 0.03.

The maximum radical concentration reached was: $[P_4 + P_4(nitroxide)] = 1.5\times10^{-4}$M.

In agreement with our EPR results, 1.4×10^{-4}M, see **Figure 2**.

After 20 days, the kinetics simulation delivered a radical concentration of 3.5×10^{-5}M in perfect agreement with our experimental results.

We can observe that the examples have quite different experimental conditions; however, they both work with the same set of rate constants.

http://dx.doi.org/10.5772/67683

In **Figure 2**, the experimental measurement of the formation and decay of the radical intermediates was represented with black square symbols and the results from the kinetic simulation process were represented with gray symbols. **Figure 4** presents the recorded EPR spectrum of the radical intermediates. Although the spectrum may fit with a nitroxide-type radical showing a hyperfine coupling constant, a_N = 12.7 G, the spectrum exhibits an increment in the central line. This central increment could be attributed to the contribution of a related allylic-type radical superimposed or to a set of related allylic-type radicals. However, the kinetic analysis generated poor concentrations for P_1, which represented the allylic radical intermediates. In addition, finally, with the aid of the computational methods, the hypothesis of the formation of cyclic nitroxides was favored [10].

2.3. (c) Theoretical evaluation of the intermediates proposed: the acyclic β-carotene neutral radicals, the acyclic nitrous β-carotene neutral radicals, and the cyclic nitroxide neutral radicals

The purpose of the following research was to unravel the structures of the related persistent intermediate or intermediates, we called them P_4 *(nitroxide)* or cyclic nitroxide neutral radical or radicals.

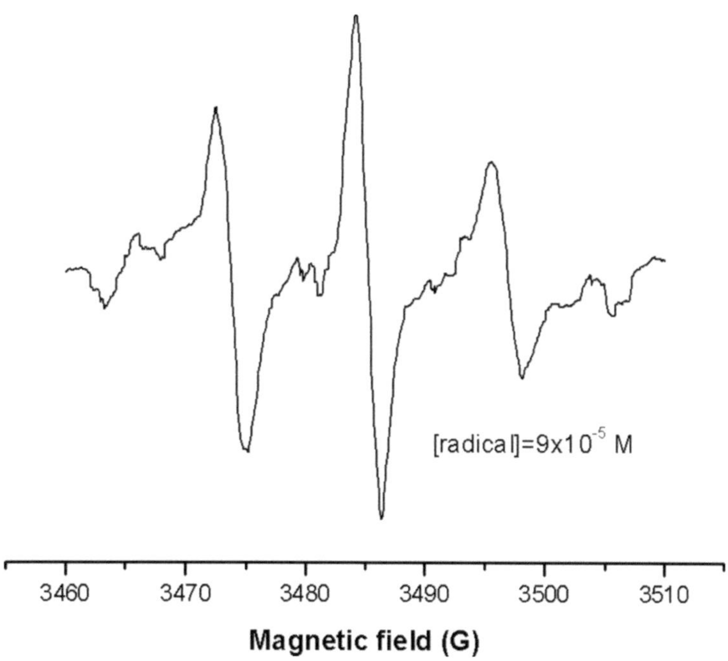

Figure 4. The figure shows a first-derivative X-band EPR spectrum of the intermediate called $P_{4(\text{nitroxide})}$, one or more type of persistent intermediate formed in the reaction of β-carotene with nitrite anion in acid medium (dioxane/water) at 298 K, under argon atmosphere. Spectrometer settings: microwave frequency, 9.92 GHz; modulation frequency, 100 kHz; microwave power, 11.3 mW; attenuation, 10 dB; field modulation, 0.125 G; receiver gain, 2×10^5; time constant, 0.05 s; scan range, 50 G; scan time, 5 s. The peak-to-peak value is of 12.7 Gauss and g = 1.994 [9].

On the basis of the kinetic studies, it was reasonable to propose intermediates of the nitroxide type. In a biological medium, one can expect that species react with each other. Or hope that radicals from β-carotene react with neighboring molecules of nitrous carotenes generating acyclic nitroxides. However, in the dilute solutions of β-carotene, where the reactions with the nitrogen oxides were carried out, the radicals kept away and intra-radical reactions would preferably take place. Thus, on that account, we considered it suitable to propose the formation of the rings.

We looked for the help of the computational and theoretical methods in order to study and compare the characteristics of the proposed structures and found out that the density functional theory (DFT) met the best conditions, given that it included the effects of electron correlation. DFT used very suitable approaches for energy exchange and for the correlation. The main conclusion was that for neutral radicals we could obtain good results with the approximate method Unrestricted Becke-style 3-parameter density functional theory using the Lee-Yang-Parr correlation functional (UB3LYP), both for the calculation of the energy of the system and for the calculation of the hyperfine coupling constants (hfccs). The selected basis sets, that used *d orbital functions*, allowed a good resolution, with an adequate cost of time, with a personal computer. The software used allowed to obtain the values of energy of each structure, which permitted to evaluate the relative stability of the radicals. Also, the software provided us the isotropic hfccs of each nucleus of the radical intermediates proposed [10].

Finally, the hfccs of the nuclei of hydrogen (^{1}H) and nitrogen (^{14}N) obtained with the theoretical methods allowed us to develop the simulation of the spectra that were experimentally recorded, like that shown in **Figure 4**.

Different possible rings with 3, 5, 6, 7, or 8 atoms were built and tested. However, only with some of them the corresponding geometric optimizations were achieved. The radicals with cycles of five atoms were found to be the most favored and those of lower energy.

In the General Reaction Pathway proposed in **Figure 3**, the following intermediates were appreciated:

- *Allylic-type radicals,* P_1
- *Nitrous allylic radicals,* P_4
- *Cyclic nitroxides formed by internal cyclization of the nitrous allylic radicals,* $P_{4(\text{nitroxide})}$

With the aim of avoiding confusion in the interpretations described, the process of formation of the radical intermediates follows.

- *Allylic-type radicals,* P_1

Figure 5 helps to visualize the possible allylic hydrogens of β-carotene (the characteristic atomic symbols (C, H) are usually not shown in this type of molecular representation). The set of neutral allylic radicals from β-carotene is symbolized by P_1 in **Figures 3** and **6**.

The molecule of β-carotene has allylic hydrogens, which are hydrogen atoms of the methyl groups at positions 5 and 5'; 9 and 9'; 13 and 13'. Allylic-type radicals are obtained by the loss, by abstraction, of those hydrogen atoms:

http://dx.doi.org/10.5772/67683

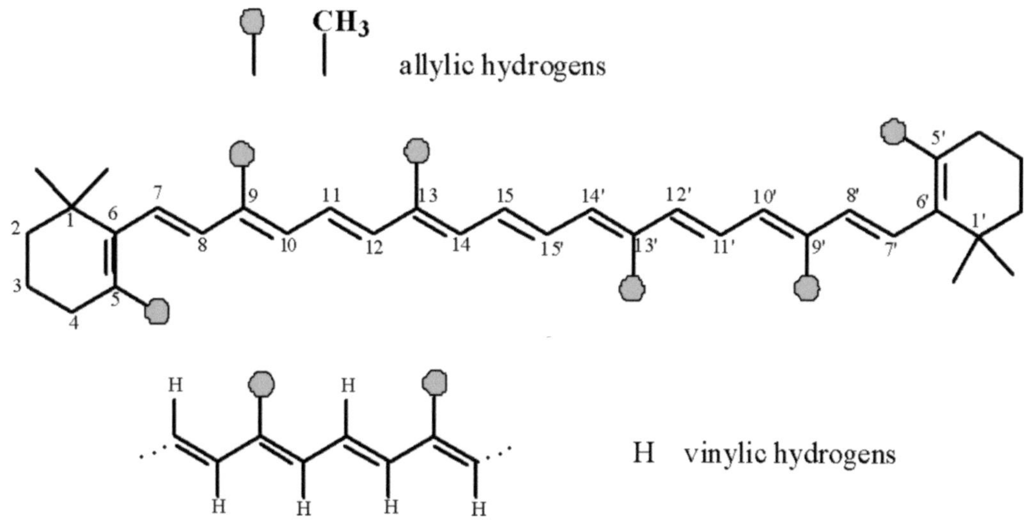

Figure 5. The molecule of β-carotene has allylic hydrogens, that are hydrogen atoms of the methyl groups at positions 5 and 5'; 9 and 9'; 13 and 13'. Vinylic hydrogen atoms are attached to the polyene chain at carbons: 7,8,10,11,12,14,15,15', 14',12',11',10',8', and 7'. Methyl groups with primary hydrogens, but not allylic type are represented with only a line or a bar in positions 1 and 1'.

$P_{10(5 \text{ or } 5')}$: The allylic-type radical is formed by the loss of a hydrogen atom of the methyl attached to carbon *5 or 5′*.

$P_{10(9 \text{ or } 9')}$: The allylic-type radical is formed by the loss of a hydrogen atom of the methyl attached to carbon *9 or 9′*.

$P_{10(13 \text{ or } 13')}$: The allylic-type radical is formed by the loss of a hydrogen atom of the methyl attached to carbon *13 or 13′*.

The molecule of β-carotene has also methyl groups (represented by a line or a bar) at positions 1 and 1'; they are primary hydrogen atoms. Those primary hydrogen atoms are not abstracted, they have higher bond energy than the allylic ones. In **Figure 5**, you can also appreciate the vinylic hydrogen atoms in the polyene chain. Those vinylic hydrogen atoms are attached to carbons: 7,8,10,11,12,14,15,15',14',12',11',10',8', and 7'; they have even higher bond energy than the primary hydrogen atoms. Observe that the radicals formed by the loss of allylic hydrogens are indeed stabilized by resonance.

In **Figure 6**, we followed the formation of the neutral carotenoid allylic radicals described below:

$P_{10(5 \text{ or } 5')}$: The β-carotene loses an allylic hydrogen atom from the methyl group attached to position 5 or 5'. The conjugation effect formed a contributing resonance structure, an allylic tertiary radical on carbon 6 (see **Figure 5**), it is $P_{11(5 \text{ or } 5')}$

$P_{12(5 \text{ or } 5')}$: It is another contributing resonance structure with the unpaired electron in carbon 13′.

$P_{10(5 \text{ or } 5')}$;$P_{11(5 \text{ or } 5')}$ and $P_{12(5 \text{ or } 5')}$ are contributing resonance structures of the *same resonance hybrid*.

$P_{10 (5 \text{ or } 5')}$

$P_{11 (5 \text{ or } 5')}$

$P_{12 (5 \text{ or } 5')}$

$P_{10 (9 \text{ or } 9')}$

$P_{11 (9 \text{ or } 9')}$

$P_{12 (9 \text{ or } 9')}$

$P_{12 (13 \text{ or } 13')}$

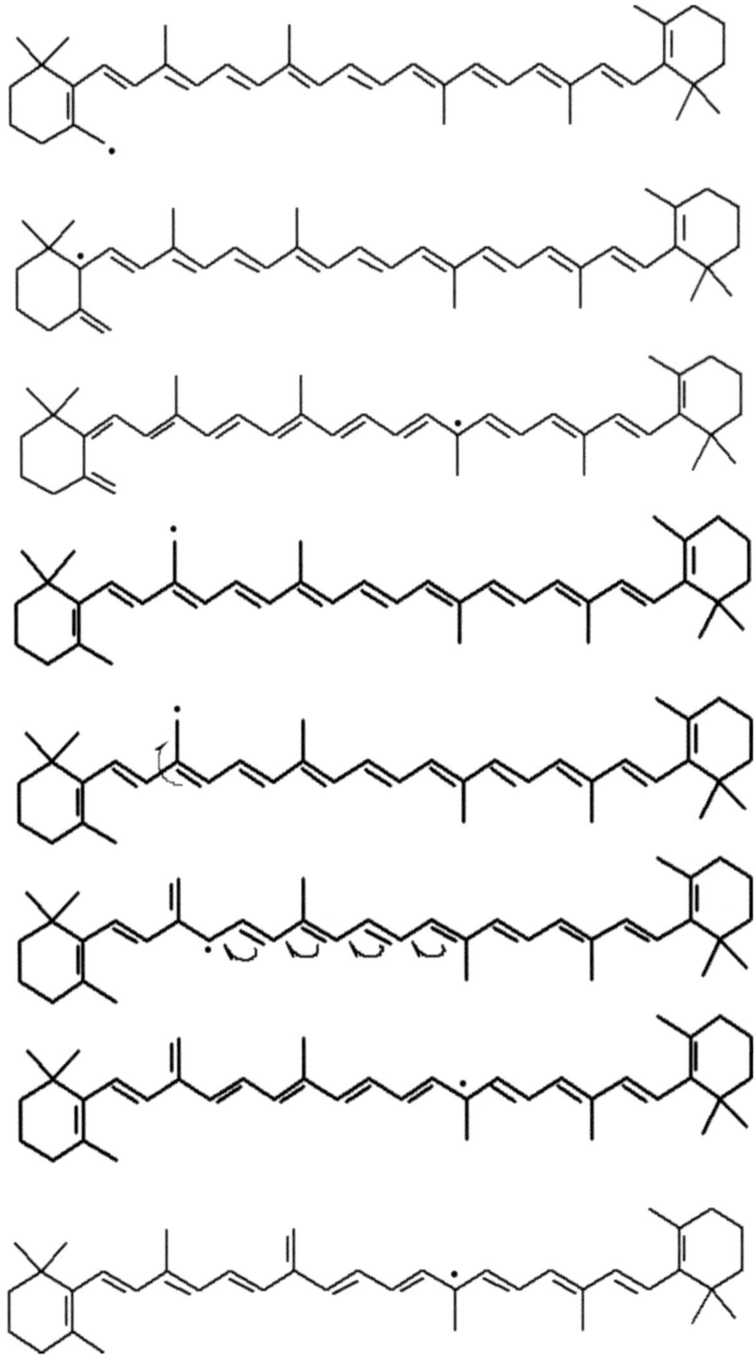

Figure 6. Several radicals of the allylic type are shown, P_1. There are three possibilities of allylic hydrogens: $P_{10 (5 \text{ or } 5')}$; $P_{10 (9 \text{ or } 9')}$, and $P_{10 (13 \text{ or } 13')}$. The second subscript indicates different contributing resonance structures, for example, $P_{11 (5 \text{ or } 5')}$ *and* $P_{12 (5 \text{ or } 5')}$. The radical formed by the loss of a methyl hydrogen atom in position 9 or 9′ is $P_{10 (9 \text{ or } 9')}$, observe that the electronic movements are shown in three related structures. $P_{1\ 0(13 \text{ or } 13')}$ is the radical formed by the loss of a methyl hydrogen atom in position 13 or 13'. For this last case, only one possible structure is designed.

http://dx.doi.org/10.5772/67683

In the case of $P_{10(9\text{ or }9')}$, the resonance or electronic movements are also shown. For $P_{10(13\text{ or }13')}$, only one structure is designed.

- *Nitrous allylic radicals, P_4*

The P_4 radicals were also investigated with the theoretical methods:

- Some of them could not reach a geometric optimization presenting convergence failure.
- Other P_4 radicals failed to reach the geometric optimization, expelling the NO group and regenerating the β-carotene molecule.
- In others, a radical was optimized geometrically but turned out to be a P_4*(nitroxide).*
- *Cyclic nitroxides formed by internal cyclization of the nitrous allylic radicals, $P_{4\text{ (nitroxide)}}$*

In **Figure 7**, we appreciate that $P_{12\text{ (5 or 5')}}$ is the intermediate that react with nitric oxide generating $P_{22'}$, the nitrous allylic compound, see **Figure 3**.

P_{22} loses by abstraction an allylic hydrogen atom from the methyl in position 9′ generating $P_{42'}$, the corresponding nitrous allylic radical and finally the formation of $P_{42\text{ (nitroxide)}}$.

(**Figure 6** in our work of 2015 has an error; the nitroxide displayed is not generated from the listed precursors. The precursors designed would actually lead to another nitroxide not shown. The optimization of that nitroxide was completed on the basis of negligible forces. The stationary point was found but the convergence criteria were reached by only three of the required four cases [10]. In the present work, we have the opportunity of showing and solving the mistake. In **Figure 7**, we appreciate the precursors that actually lead to P_{42}*(nitroxide)*.)

$P_{12\ (5\text{ or }5')}$

P_{22}

P_{42}

P_{42} *(nitroxide)*

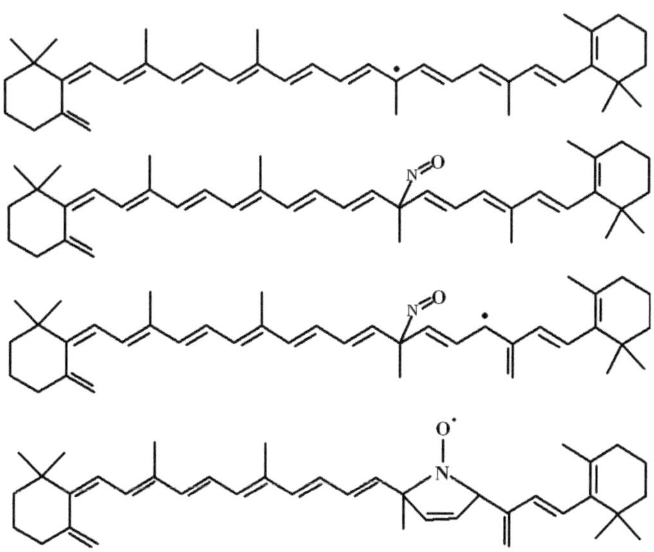

Figure 7. Formation of the cyclic P_{42}*(nitroxide)* from the allylic precursor. The tertiary allylic radical $P_{12\text{ (5 or 5')}}$ is a contributing resonance structure of high weight.

In **Figure 8**, the optimized structure of $P_{42\ (\text{nitroxide})}$ is displayed. Each atom is labeled with a number.

In order to carry out the simulation, three selected nitroxides were chosen, those of lower energy and those that best met the requirements for optimization. We selected two rings of five atoms ($P_{41\ (\text{nitroxide})}$; $P_{42\ (\text{nitroxide})}$) and one of eight atoms ($P_{43\ (\text{nitroxide})}$).

We simulated quite satisfactorily the experimental recorded EPR spectrum in **Figure 4**, by adding the theoretical spectra of $P_{42\ (\text{nitroxide})}$ from $P_{12\ (5\ \text{or}\ 5')}$ and $P_{41\ (\text{nitroxide})}$ from $P_{11\ (5\ \text{or}\ 5')}$, both cyclic nitroxides of five atoms. Also we built another simulation by adding $P_{42(\text{nitroxide})}$, $P_{41\ (\text{nitroxide})}$, and $P_{43\ (\text{nitroxide})}$ from $P_{10\ (9\ \text{or}\ 9')}$, a cyclic nitroxide of eight atoms [10].

Figure 9 shows the theoretical spectra and simulation of the persistent EPR recorded spectrum of **Figure 4**. In part (a), we may appreciate the theoretical EPR spectra of the nitroxides P_{41}*(nitroxide)*, P_{42}*(nitroxide)*, *and* P_{43}*(nitroxide)*. In order to simulate the experimental recorded spectra, one must add up the spectra of nitroxides, multiplying the values of P_{41}*(nitroxide)* by the factor 1 and P_{42}*(nitroxide)* by the factor 0.4 and multiply the values of P_{43}*(nitroxide)* by the factor 0.2. Part (b) shows the sum that simulates quite well the experimental spectrum shown in **Figure 4**.

Simulations were also carried out considering the contribution of the β-carotene neutral radicals or allylic neutral radicals. It was observed that with less than 0.1 order factors, there were no appreciable changes, as is expected due to the result of the kinetic modeling. Although the β-carotene neutral radicals could be persistent, the nitroxide-type signal obtained would not be modified.

The theoretical spectra were built from the calculated isotropic hyperfine coupling constants. The hfccs values larger than 0.4 Gauss are shown in **Table 3**.

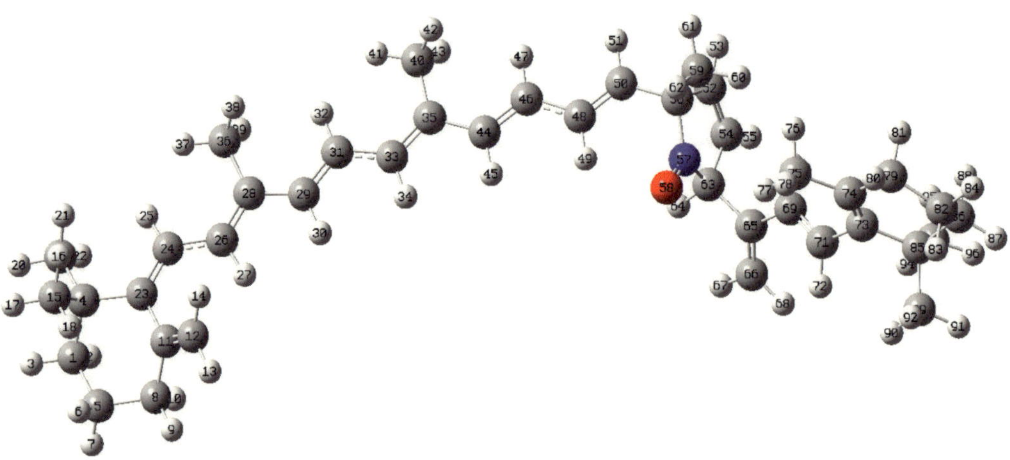

Figure 8. Visualization of the optimized geometric structure of the intermediate radical proposed in **Figure 7**: P_{42}*(nitroxide)*.

http://dx.doi.org/10.5772/67683

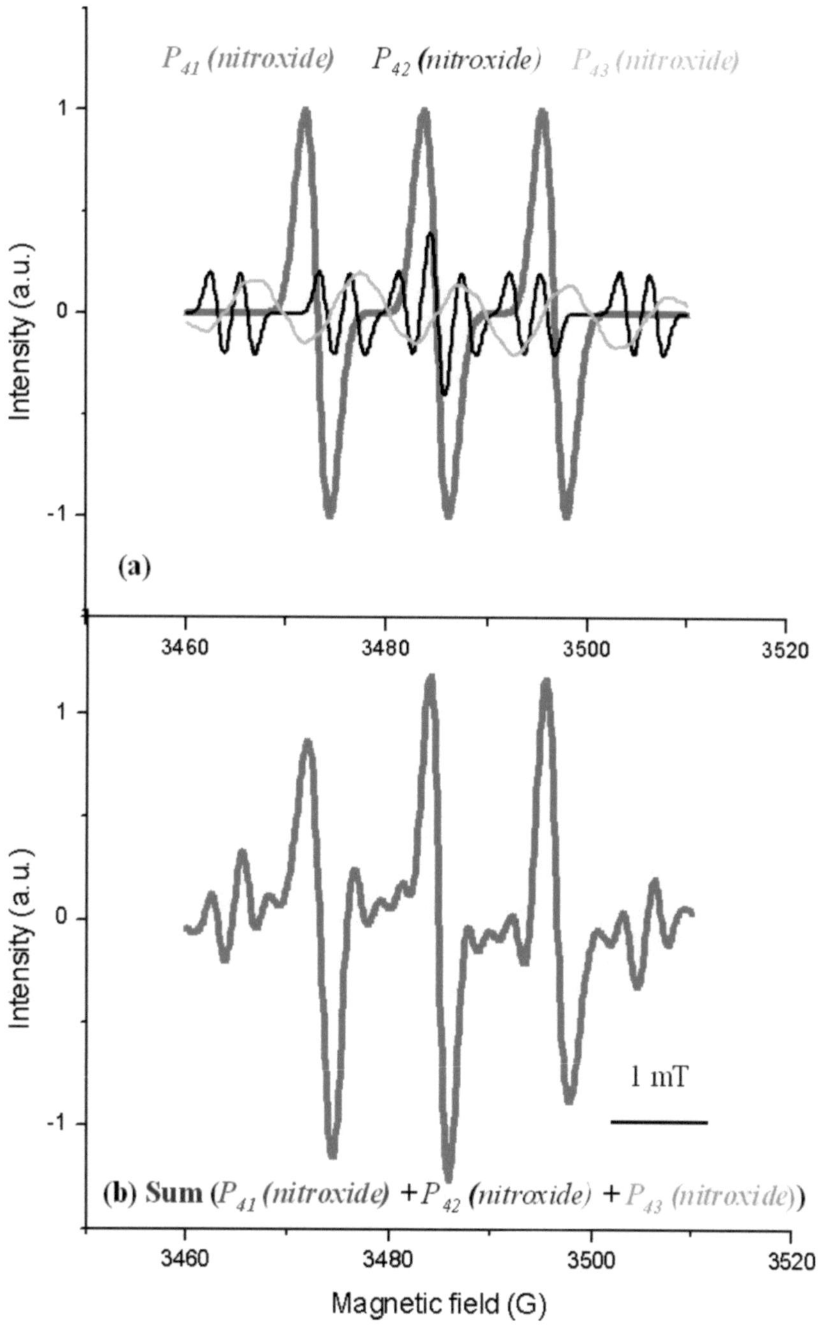

Figure 9. EPR theoretical spectra and simulation. In the part (a) the EPR theoretical spectra of the nitroxides are recorded: ***P41(nitroxide)*** *in gray*, P_{42}*(nitroxide)* in black, and ***P43(nitroxide)*** in pale gray color. The values of ***P41(nitroxide)*** are multiplied by a factor of 1, P_{42}*(nitroxide)* are multiplied by a factor of 0.4 and the values of ***P43(nitroxide)*** by a factor of 0.2. The part (b) shows the sum of the spectra in order to simulate the experimental spectrum. The peak-to-peak value is 12.1 Gauss, (10 Gauss ≡ 1 mT). Data from **Table 3** are used.

Nuclei	P_{41}*(nitroxide)*	Nuclei	P_{42}*(nitroxide)*	Nuclei	P_{43}*(nitroxide)*
3 H	0.73171	49 H	0.42705	28 H	0.94710
10 H	0.91335	51 H	2.99147	31 H	0.45231
14 H	0.56426	53 H	−0.53599	53 H	2.00175
18 H	0.57683	55 H	−0.55714	54 H	9.42852
25 H	−0.53235	57 **N**	10.95771	56 H	19.03785
27 H	−0.58125	60 H	−0.45979	57 H	2.93589
29 **N**	11.68685	62 H	−0.60386	88 **N**	11.40254
33 H	−0.57914	64 H	18.74614	92 H	2.15932
34 H	−0.47031			96 H	1.59845
36 H	−0.42026				

[a]The hfccs values were obtained by applying the method of calculation B3LYP/6-31G (d) to the optimized geometries. The hfccs values larger than 0.4 Gauss are shown. Gaussian offers results in Gauss, 10 G ≡ 1 mT.

P_{41}*(nitroxide)* and P_{42}*(nitroxide)* are rings of five atoms. P_{43}*(nitroxide)* corresponds to a ring of eight atoms.

Each atom is labeled with a number, as in **Figure 8** for P_{42}*(nitroxide)*.

Table 3. Isotropic hyperfine coupling constants (hfccs), cyclic P_4*(nitroxides)*, (Gauss)[a].

3. Conclusions and perspective

We have provided considerable information about the behavior of nitrogen dioxide in some solvents. Also, we followed the action of both NO_2 and NO over some organic compounds. Most relevant was the study of the reaction of both oxides with the β-carotene molecule.

It is important to highlight that two spectroscopic techniques were used: UV-vis and EPR. UV-vis is fundamental for the tracking of reagents, and EPR is essential for the monitoring of the radical intermediates.

Finally, we used the theoretical and computational methods that provided support to the characteristics of the intermediates. Calculations for β-carotene cyclic nitroxide intermediates, β-carotene allyl radicals, and nitrous β-carotene radical showed that the combination of the 6-31G(d,p) basis set and the B3LYP exchange-correlation functional provided a quite accurate description of the hfccs of the nuclei of hydrogen and nitrogen. The calculation of the hfcc for nitrogen nuclei (^{14}N, nucleus) in the case of the nitroxide persistent radicals was particularly difficult. This problem with the nitroxides was investigated and finally deduced that the choice of the basis sets was very important. Especially the number and nature of *d orbital functions* must be taken into account [10].

β-Carotene or other carotenoids compounds would be in the lipid and anaerobic part of cells where the investigated reactions could take place. The β-carotene could be a convenient scavenger, acting as a protective agent in biological media. The cyclic persistent nitroxides may react by disproportionation giving rise to hydroxylated carotenoid compounds and nitrones.

Hydroxylated compounds can pass into the lymphatic system and be replaced by new molecules of β-carotene. The nitrone is a new scavenger that will lead to a continuously replacement process [10]. It is important to remember that the persistent nitroxides are still present in solutions after 20 days with a concentration of $3.5 \cdot 10^{-5}$ M [9].

On the other hand, the nitroxides may react differently. For example:

(a) β-Carotene may not be a good radical trapping antioxidant but the nitroxide may be [21].

(b) The combination of the carotenoids lutein or zeaxanthin and lipid-soluble nitroxides exerted strong synergistic protection against singlet oxygen-induced lipid peroxidation. The synergistic effect may be explained in terms of protection of the intact lutein or zeaxanthin structure by effective scavenging of free radicals by nitroxides [22].

Moreover, we can see that nitric oxide is generated in different types of cells. Neuronal nitric oxide synthase (nNOS) is constituvely expressed in specific neurons of the brain. In addition to brain tissue, nNOS has been identified in adrenal glands, in epithelial cells of various organs. In mammalians, the largest source of nNOS in terms of tissue mass is in the skeletal muscle [23].

It might be expected that nitrogen oxides and non-polar carotenoids compounds have together an important biological function. Take into account that in carotenoids with a hydroxyl group, the reaction path may be surely different.

The isolation of the cyclic nitroxides described would also be very important.

Author details

Sara N. Mendiara[1*] and Luis. J Perissinotti[1,2]

*Address all correspondence to: mendiara@gmail.com

1 Department of Chemistry, Faculty of Exact and Natural Sciences, National University of Mar del Plata, Mar del Plata, Buenos Aires, Argentina

2 Commission of Scientific Research of the Province of Buenos Aires (CIC), La Plata, Buenos Aires, Argentina

References

[1] Mendiara S, Sagedahl A, Perissinotti L. An electron paramagnetic resonance study of nitrogen dioxide dissolved in water, carbon tetrachloride and some organic compounds. Appl. Magn. Reson. 2001; 20: 275–287. doi:10.1007/BF03162326

[2] Halliwell B, Gutteridge JMC. Free radicals in biology and medicine, 4th ed. New York: Oxford University Press; 2007. ISBN 978-0-19-856868-1

[3] Möller M, Li Q, Lancaster Jr. J, Denicola A. Acceleration of nitric oxide autoxidation and nitrosation by membranes. IUBMB Life. 2007; 59: 243–248. doi:10.1080/15216540701311147

[4] Gruszecki W, Strzałka K. Carotenoids as modulators of lipid membrane physical properties. Biochim. Biophys. Acta. 2005; 1740: 108–115. doi:10.1016/j.bbadis.2004.11.015

[5] Rao A, Rao L. Carotenoids and human health. Pharmacol. Res. 2007; 55: 207–216. doi:10.1016/j.phrs.2007.01.012

[6] Kopec R, Schwartz S. Carotenoid cleavage dioxygenase and presence of apo-carotenoids in biological matrices. In: Winterhalter P, et al. (eds) Carotenoid Cleavage Products. ACS Symposium Series. Washington, DC: American Chemical Society; 2013. pp. 31–41. doi:10.1021/bk-2013-1134.ch004

[7] Johnson E, Vishwanathan R, Johnson M, Hausman D, Davey A, Scott T, Green R, Miller L, Gearing M, Woodard J, Nelson P, Chung H, Schalch W, Wittwer J, Poon L. Relationship between serum and brain carotenoids, α-tocopherol, and retinol concentrations and cognitive performance in the oldest old from the Georgia centenary study. J. Aging Res. 2013; 2013: 7–10. doi:10.1155/2013/951786

[8] Hammond B, Jr. Dietary carotenoids and the nervous system. Foods. 2015; 4: 698–701. doi:10.3390/foods4040698

[9] Mendiara S, Baquero R, Katunar M, Mansilla A, Perissinotti L. Reaction of β-carotene with nitrite anion in a homogeneous acid system. An electron paramagnetic resonance and ultraviolet-visible study. Appl. Magn. Reson. 2009; 35: 549–567. doi:10.1007/s00723-009-0185-1

[10] Mendiara S, Perissinotti L. Neutral radicals in the reaction of β-carotene with NO_2 and NO: computational study and simulation of the EPR spectra. Appl. Magn. Reson. 2015; 46: 1301–1322. doi:10.1007/s00723-015-0732-x

[11] Gabr I, Patel R, Symons M, Wilson M. Novel reactions of nitric oxide in biological systems. J. Chem. Soc. Chem. Commun. 1995; 1995: 915–916. doi:10.1039/C39950000915

[12] Wolak M, Stochel G, Hamza M., van Eldik R. Aquacobalamin (Vitamin B_{12a}) does not bind NO in aqueous solution. Nitrite impurities account for observed reaction. Inorg. Chem. 2000; 39: 2018–2019. doi:10.1021/ic991266d

[13] Brown J, Jr. The reaction of nitric oxide with isobutylene. J. Am. Chem. Soc. 1957; 79: 2480–2488. doi:10.1021/ja01567a035

[14] Rockenbauer A, Korecz L. Comment on conversion of nitric oxide into a nitroxide radical using 2,3-dimethylbutadiene and 2,5-dimethylhexadiene. J. Chem. Soc. Chem. Commun. 1994; 1994: 145. doi:10.1039/C39940000145

[15] Mendiara S, Perissinotti L. Dissociation equilibrium of dinitrogen tetroxide in organic solvents: an electron paramagnetic resonance measurement. Appl. Magn. Reson. 2003; 25: 323–346. doi:10.1007/BF03166693

[16] Cundall, J. Dissociation of liquid nitrogen peroxide. Part II. The influence of the solvent. J. Chem. Soc. Trans. 1895; 67: 794–811. doi:10.1039/CT8956700794

[17] Gray, P., Rathbone, P. Dissociation of liquid dinitrogen tetroxide; Henry's law coefficients and entropies of solution, and the thermodynamics of homolytic dissociation in the pure liquid. J. Chem. Soc. 1958; 1958: 3550–3557. doi:10.1039/JR9580003550

[18] Redmond T, Wayland B. Dimerization of nitrogen dioxide in solution: a comparison of solution thermodynamics with the gas phase. J. Phys. Chem. 1968; 72: 1626–1629. doi:10.1021/j100851a040

[19] Moore J, Pearson R. Kinetics and Mechanism, 3rd ed. New York: John Willey & Sons; 1981. 240 p. ISBN 0-471-03558-0

[20] Geddes J. The fluidity of dioxane-water mixtures. J. Am. Chem. Soc. 1933; 55: 4832–4837. doi: 10.1021/ja01339a017

[21] Ingold K, Derek A. Advances in radical-trapping antioxidant chemistry in the 21stTable century: a kinetics and mechanisms perspective. Chem. Rev. 2014; 114: 9022–9046. doi:10.1021/cr500226n

[22] Zareba M, Widomska J, Burke J, Subczynski W. Nitroxide free radicals protect macular carotenoids against chemical destruction (bleaching) during lipid peroxidation. Free Radic. Biol. Med. 2016; 101: 446–454. doi:10.1016/j.freeradbiomed.2016.11.012

[23] Förstermann U, Sessa W. Nitric oxide synthases: regulation and function. Eur. Heart J. 2012; 33: 829–837. doi:10.1093/eurheartj/ehr304

Chapter 6

Influence of Environmental Stress toward Carotenogenesis Regulatory Mechanism through *In Vitro* Model System

Rashidi Othman, Norazian Mohd Hassan and
Farah Ayuni Mohd Hatta

Additional information is available at the end of the chapter

http://dx.doi.org/10.5772/67724

Abstract

Carotenoid biosynthesis is influenced by some aspects and is liable to geometric isomerisation with the existence of oxygen, light, and heat, which affect color degradation and oxidation. The major problems related to carotenoid accumulation inherently originate from pigment instability. This chapter discusses an overview on the influence of stringent control of genetic, developmental, and environmental factors toward carotenoid biogenesis in potato minitubers through the potential model system for rapid initiation, extraction, and analysis of carotenoids. The outcome of this experimental system is a discovery of variables regulating carotenoid accumulation as a result of the environmental change assessment through manipulation of drought stress, light intensity, and nutrient strength on carotenoid accumulation.

Keywords: carotenogenesis, environmental stress, *in vitro*, model system, elicitors

1. Introduction

Considerable research interest has recently focused on the improvement of both transgenic and conventional propagation techniques to enrich total and individual carotenoid composition in potatoes [1–4]. Unfortunately, little information is available on the influence of the environment on the carotenoid content in potatoes, especially growing seasons and locations. Genotype and environment interactions have been reported to account for alteration in free amino acids, protein, and sugar composition [5–13]. In addition, the total glycoalkaloid content of potato tubers was found greatly affected as a result of environmental changes during the growing season [14], even though there are also strong genetic effects [15, 16]. Seasonal differences, growing conditions, locations, genotypes, and postharvest storage conditions are among the factors that can be significantly affecting the quality and nutritional value of potatoes [17–20]. The bioavailability of carotenoids is characterized as an intricate issue and influenced by various factors [21]. In our study of interseasonal and genotype interactions, the data revealed that variations in total carotenoid content and the concentration of individual carotenoid pigments are due to the strong relationship between genotype and growing seasons. This assumption is supported by Chloupek and Hrstkova [22] in their observations of 26 crops over a 43-year period growing seasons; where yield adaptability over time was controlled largely by weather and small variations from year to year in agronomical practices. In other words, major factors influencing yield are location, year, and their interactions. They also observed that yield variation of the 26 crops, including potato, in the Europe was greater than in the USA by nearly two times. In another case, the level of polyphenols in potatoes has been reported to have significant difference with environmental conditions and genetics [22]. A strong relationship and association between growing location of potato, the yellow color intensity in tuber flesh, and its total carotenoid content have also been reported. Report [1] demonstrated that environmental factors may affect on the yellow intensity of tuber flesh. The correlation between genotypes and environment can be indicative of the particular potato cultivar for best adapted to certain locations. For example, in 2004/2005 growing season in New Zealand, Agria was found to have a substantially higher carotenoid content relative to other cultivars with mostly lutein and no zeaxanthin, whereas in 2006/2007 Agria contained all five carotenoids with relatively high concentration of zeaxanthin [23]. A notable difference between the two seasons was the accumulation of zeaxanthin in 2006/2007 and the absence of zeaxanthin in 2004/2005.

pH can affect epoxidation and de-epoxidation reactions in the xanthophyll cycle [24]. Hydroxylation can convert α- and β-carotene to lutein and zeaxanthin, respectively. Violaxanthin is formed from zeaxanthin due to epoxidation and de-epoxidation that can transform violaxanthin back to zeaxanthin. This reaction sequence is reversible and mediated by pH [25, 26]. Epoxidation will occur in darkness or under low light condition and activity is optimal near pH 7.5 [27, 28], whereas de-epoxidation activity is active at pH below 6.5 and optimal at approximately pH 5.2 [24].

Zeaxanthin happens only in trace amounts under physiological conditions *in vivo* or without stress condition [29–31]. Nevertheless, zeaxanthin occurs upon de-epoxidation through the reversible xanthophyll cycle operation due to exposure under irradiance stress or high light condition [32, 33]. Although zeaxanthin accumulates during irradiance stress, that association is normally only transient. Upon recovery under low light or in darkness, zeaxanthin will disappear [33]. In addition, it was revealed from recent *in vitro* studies [34, 35], upon analysis of zeaxanthin accumulating *Arabidopsis thaliana* mutants [36–38] and from the green alga *Scenedesmus obliquus* [39], that zeaxanthin could replace lutein and violaxanthin under irradiance stress.

There are two possibilities to explain the accumulation of zeaxanthin in 2006/2007 season and not previous season:

(i) The conversion of individual carotenoids such as violaxanthin, neoxanthin to zeaxanthin is due to irradiance stress condition from high light exposure. This will promote the conversion of other carotenoids to zeaxanthin from the β-carotene and α-carotene branch point. As a result, zeaxanthin concentration will increase. Consequently, the precursor supply for ABA biosynthesis and the plant responds will inhibit as the carotenogenic metabolic flux increases to compensate for this restriction [29, 40]. Previous study [41] also reported that lutein and violaxanthin can transform to zeaxanthin successfully under irradiance stress condition.

(ii) The presence and absence of zeaxanthin are in response to alterations in pH. Acidity will trigger the de-epoxidation reaction by the conversion of violaxanthin and other precursors of ABA to zeaxanthin, whereas alkaline conditions will induce lutein or the supply of precursors for ABA biosynthesis, which will lead to the conversion of zeaxanthin to violaxanthin, neoxanthin or other precursors for ABA biosynthesis through epoxidation reaction [42].

Overall, this study clearly demonstrated that the total and the individual pigment content of carotenoids in potato tubers were depending on the growing season, subsequently, affect their quality and nutritional content. Thus, along with genotypic factors, environmental factors also take on an important part in regulating the accumulation of individual carotenoids in potato tubers, especially in Agria and Desiree. Between seasons, lutein has been transformed into zeaxanthin in Agria, whereas neoxanthin has been transformed into zeaxanthin in Desiree. These findings evidently suggest that selection of high or low carotenoid tuber levels cannot be established on the basis of a single year's results. Still, valid comparisons can be established between data from different years if the material is stored and developed under similar environmental conditions. This study suggests that environmental factors such as seasonal climatic variation may influence the accumulation of potato tuber carotenoids content and composition. Apparently, further research using potato plant materials produced under different environmental conditions are needed to support this theory.

2. Experimental design

2.1. Tissue culture and minituber initiation

Virus free *in vitro* plants of cultivars Agria and Desiree were supplied by the New Zealand Institute for Crop & Food Research Ltd. These were cultured in a incubation room at 24°C day and night temperature, with a 16-h photoperiod under cool white fluorescent light at 80–85 μmol m^{-2} s^{-1}. Every 4 weeks, the *in vitro* plants were subcultured as nodal cuttings on potato multiplication medium (PMM) composed of Murashige and Skoog (MS) salts and vitamins [43] added with 30 g/L sucrose, 40 mg/L ascorbic acid, 500 mg/L casein hydrolysate, and 10 g/L agar in accordance with the procedure of Conner et al. [44]. Media was adjusted to pH 5.7 and sterilized by autoclaving (15 min, 121°C) and 50 ml aliquots poured into presterilized 290 ml plastic bottles (80 mm diameter × 60 mm high; Vertex Plastics, Hamilton, New

Zealand). For minituber initiation, individual shoots of 3–4 nodes from vigorously growing 4-week-old cultures were transferred into 40 ml of liquid tuber initiation medium (TIM) in 250 ml polycarbonate culture vessels (7 cm diameter × 8 cm high). The TIM contained the same constituents as PMM, except with the addition of 80 g/L sucrose, 5 mg/L benzyladenine, 2.5 mg/L ancymidol, and no agar. Nine shoots were placed upright into each culture vessel and were incubated in darkness at 25°C. Minitubers were classified as such when their diameter exceeded 2 mm and normally grew up to more than 5 mm diameter within 4 weeks.

2.2. Effect of environmental factors on carotenoid biosynthesis

In three independent experiments, the influence of light, water stress, and nutrient availability on carotenoid biosynthesis were tested in both Agria and Desiree. Minitubers harvested after 4 weeks from two culture vessels were pooled for each of three replicates established under the following conditions:

1. Light versus darkness by incubation under cool white fluorescent light (80–85 μmol m^{-2} s^{-1}, 16 h photoperiod) with dark condition imposed by carefully wrapping the culture vessels in aluminium foil.
2. Incubation in darkness with and without 50 mM PEG 4000 to impose water stress.
3. Incubation in darkness at three concentrations of MS salts (one tenth, half, and full strength).

2.3. Minituber extraction and analysis of carotenoids

Minitubers were harvested and pooled for each replicated treatment, cut in half, and freeze-dried as combined skin and flesh samples for 7 days. The samples were then ground into fine powder and kept at −80°C until further analysis.

The extraction procedure followed the methods described in several reports [45–47]. 0.1 g of each powdered sample was rehydrated with distilled water and extracted with a mixture of acetone and methanol (7:3) at room temperature until colorless. The crude extracted was then centrifuged for 5 min at 10,000 g and stored in darkness at 4°C until analysis. The same volume of hexane and distilled water was added to the combined supernatants to extract carotenoids. The mixture was set aside until separation occurred and the upper layer holding the carotenoids was collected. The upper hexane layer was then removed using a gentle stream of oxygen-free nitrogen until the collected carotenoid was dried completely.

The carotenoids HPLC analysis was performed on an Agilent model 1100 series equipped with a binary pump, autosampler injector, micro vacuum degassers, thermostatted column compartment, and a diode array detector [45]. The column used was a Luna C18 end capped 5 μm, 250 × 4.6 mm reverse phase column (Phenomenex Auckland, New Zealand). The solvents used were (A) acetonitrile: water (9:1, v/v) and (B) ethyl acetate. The gradient of solvent used was developed as follows: 0–40% solvent B (0–20 min), 40–60% solvent B (20–25 min), 60–100% solvent B (25–25.1 min), 100% solvent B (25.1–35 min), and 100–0% solvent B (35–35.1 min) at

http://dx.doi.org/10.5772/67724

1.0 ml min^{-1} flow rate. The column temperature was maintained at 20°C and was allowed to reequilibrate in 100% solvent A for 10 min prior to the next injection. The volume of an injection was 10 µL. Carotenoid standards β-carotene, violaxanthin, lutein, and neoxanthin were isolated from *Eruca sativa* (roquette or rocket salad) by open column chromatography [48], whereas zeaxanthin was obtained commercially from Sigma-Aldrich (Auckland, New Zealand).

3. Results

3.1. Effect of light on carotenoid accumulation in potato minitubers

Statistical analysis demonstrated that there was a highly significant difference ($P < 0.0001$) in carotenoid content in Agria minitubers developing in the dark and light. Agra minitubers accumulated four individual carotenoid compounds (violaxanthin, zeaxanthin, lutein, and β-carotene) when developing in both dark and light. The two predominant carotenoids were violaxanthin and zeaxanthin. Neoxanthin was not detectable in either dark or light treatments. However, development of Agria minitubers in light resulted in an approximate doubling of the total carotenoid content compared minitubers developing in darkness (**Figure 1**). The amount of each individual carotenoid also approximately doubled upon development in light, especially for violaxanthin and zeaxanthin. Analysis of variance comparing Desiree minitubers grown in the dark and light also exhibits highly significant differences ($P < 0.0001$) in carotenoid content. As shown in **Figure 1**, five individual carotenoids (neoxanthin, violaxanthin, zeaxanthin, lutein, and β-carotene) were found in Desiree minitubers grown in darkness, but upon development in light only four (neoxanthin, violaxanthin, lutein, and β-carotene) were detected, with an absence of zeaxanthin. After development in light, total carotenoid content approximately doubled and reflected an increase in neoxanthin and violaxanthin.

3.2. Effect of PEG on carotenoid accumulation in potato minitubers

Analysis of variance revealed that there was a highly significant difference ($P < 0.0001$) in carotenoid content in response to the water stress treatment during development of Agria minitubers. Agria minitubers developing in the presence of PEG (**Figure 2**) exhibit an increased total carotenoid content. This increase reflected a substantially higher amount of violaxanthin and occurred despite the total absence of zeaxanthin in the presence of PEG. Analysis of variance also indicates highly significant differences ($P < 0.0001$) in carotenoid content for Desiree minitubers developing in the presence of water stress. As shown in **Figure 2**, total carotenoid content increased in minitubers developing in the PEG treatment. This reflected an increase in both neoxanthin and violaxanthin, with traces of lutein being observed in both treatments.

3.3. Effect of nutrient stress on carotenoid accumulation in potato minitubers

Nutrient stress during Agria minituber development resulted in a highly significant difference ($P < 0.0001$) in carotenoid content. When MS salt strength increased from 0.1× to 0.5×, total carotenoid, violaxanthin, and β-carotene content decreased, accompanied by a slight increase in lutein concentration. However, when MS salt strength increased from 0.5× to 1.0×, total carotenoid, violaxanthin, and β-carotene increased, whereas lutein concentration decreased (**Figure 3**).

Analysis of variance also establishes highly significant differences ($P < 0.0001$) in carotenoid content in Desiree minitubers developing in varying MS salt strengths. As shown in **Figure 3**, when MS salt strength increased from 0.1× to 0.5×, total carotenoid content slightly increased

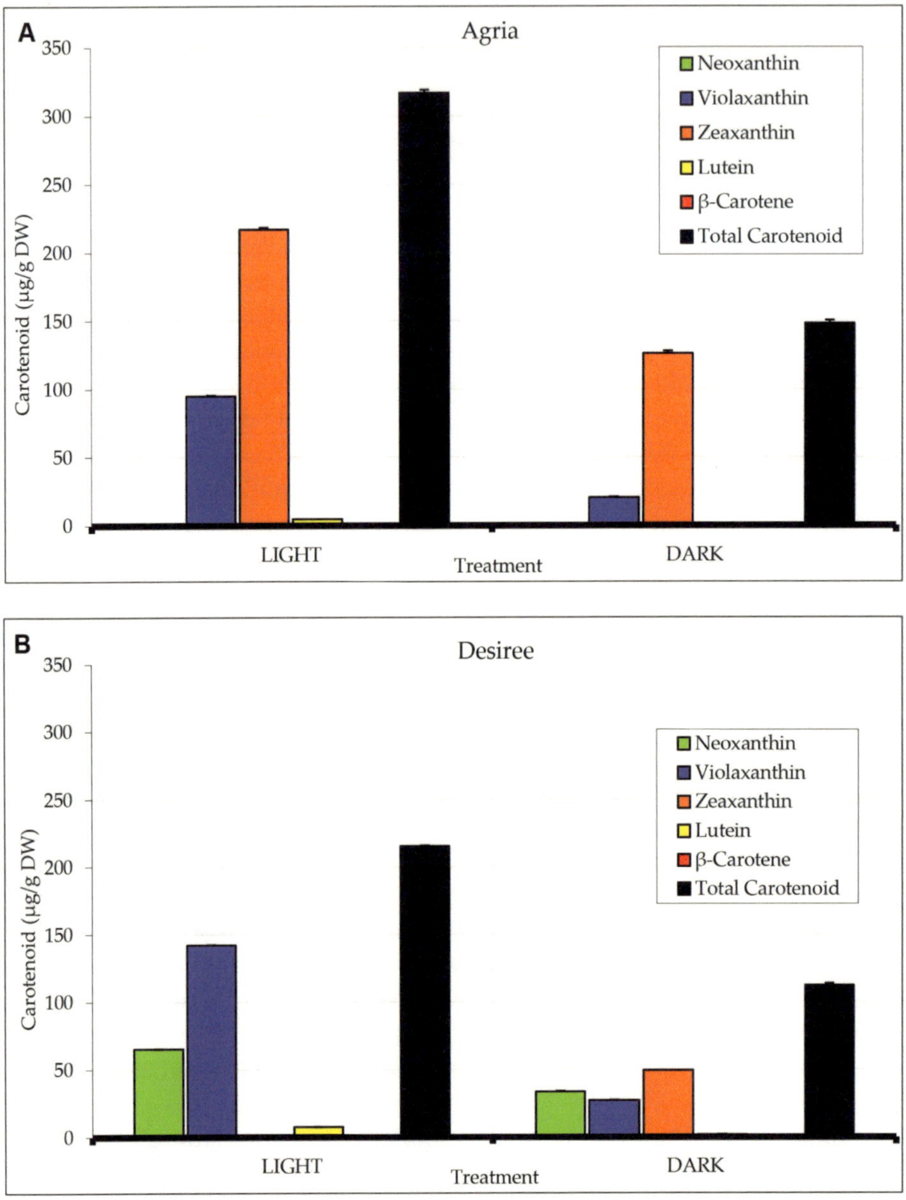

Figure 1. Analysis of carotenoid content (µg/g DW) of Agria and Desiree minitubers in response to light; (A) individual and total carotenoid content (µg/g DW) of Agria minitubers developing in light and dark; (B) individual and total carotenoid content (µg/g DW) of Desiree minitubers developing in light and dark; error bars represent ± SE.

http://dx.doi.org/10.5772/67724

due to minor changes in neoxanthin and lutein. In contrast, upon further increases in MS salt strength, 0.5–1.0×, total carotenoid content and individual carotenoids, especially neoxanthin, violaxanthin, and lutein, decreased. No changes were observed in β-carotene when MS salt strength increased from 0.1× to 0.5× for the development of Desiree minitubers.

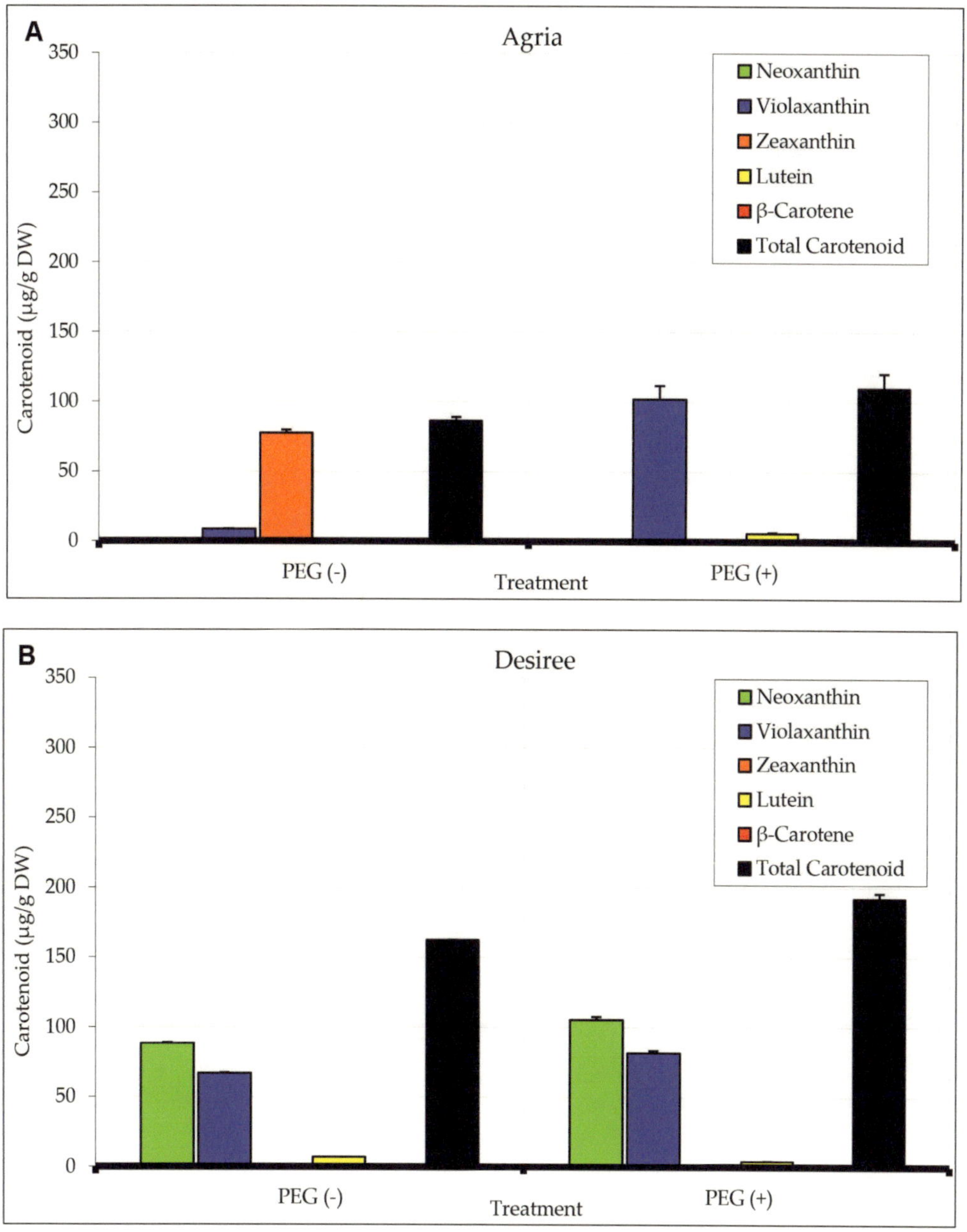

Figure 2. Analysis of carotenoid content (μg/g DW) of Agria and Desiree minitubers in response to water stress; (A) individual and total carotenoid content (μg/g DW) of Agria upon development with and without PEG treatment; (B) individual and total carotenoid content (μg/g DW) of Desiree upon development with and without PEG treatment; error bars represent ± SE.

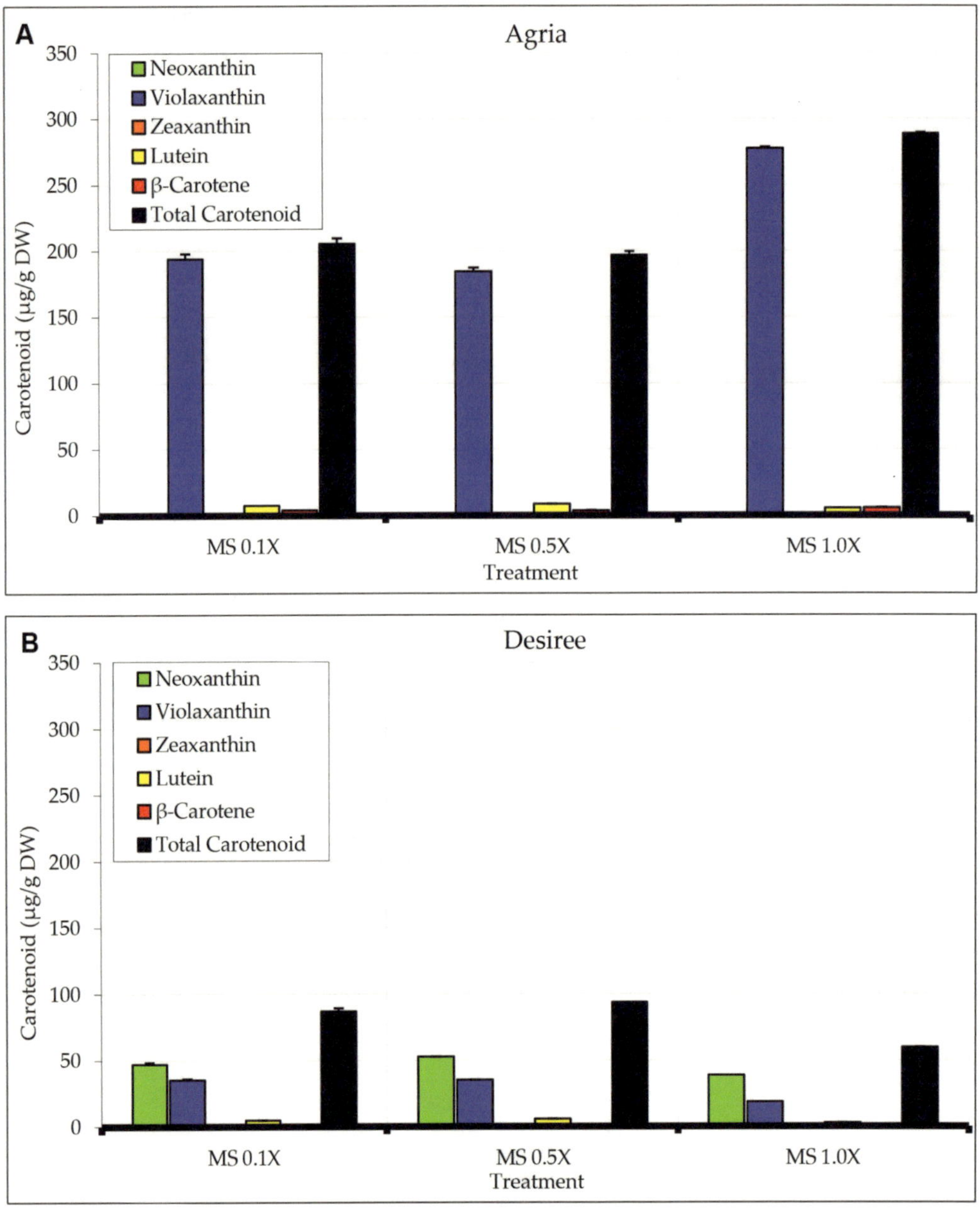

Figure 3. Analysis of carotenoid content (μg/g DW) of Agria and Desiree minitubers in response to nutrient levels; (A) individual and total carotenoids content (μg/g DW) of Agria upon 0.1×, 0.5× and 1.0× MS salt stress; (B) individual and total carotenoids content (μg/g DW) of Desiree upon 0.1×, 0.5× and 1.0× MS salt stress; error bars represent ± SE.

4. Discussion

The development of potato minitubers through *in vitro* system has proved to be an effective experimental system for investigating the environmental factors involved in regulating the

http://dx.doi.org/10.5772/67724

carotenoid biosynthesis. This potential model system has been used because of several advantages compared to the field cultivated tubers:

(i) Rapid initiation of minitubers within four weeks from establishing the experiment rather than whole growing season in the field;

(ii) The environmental conditions are easy to control because of the small size of the plantlets;

(iii) Potato minitubers were easily exposed to different types of environmental treatments effect;

(iv) Variation between tubers was minimized; and

(v) Extraction and analysis of carotenoids can be done by using potato minitubers.

Environmental stress is described as external conditions that adversely affect growth, development, or productivity [49]. Plants respond to stress by various means such as transformed gene expression, trigger cellular metabolism, and variations in growth rates and crop yields. There are two categories of stress:

(i) Biotic – caused by other organisms; and

(ii) Abiotic – resulted from the physical or chemical environment excess or deficiency.

Abiotic or physical and chemical environmental conditions can trigger stress and regulate the carotenoid biosynthesis and of this light, water stress, and nutrient are among the important factors. Species, genotype, and development age are factors influencing the resistance or sensitivity of plants to stress. Three stress resistance mechanisms are as follows:

(i) Avoidance mechanisms – prevents exposure to stress;

(ii) Tolerance mechanisms – permit the plant to withstand stress; and

(iii) Acclimation – modify their physiology in response stress.

Plants cultivated in full sunlight exposure usually receive and absorb more light for photosynthesis process. Carotenoids have a significant role to protect photosynthetic organisms against excessive light [50, 51] and these functions have been proven *in vitro* in photosystem II complexes [52]. Light is a major stress factor in plants causing photoinhibition and photooxidation in photosynthetic tissues and inhibit productivity. Light is also one of the main factors regulating carotenoid biosynthesis [53]. Current researches have reported a clear relationship between the dissipation of excess excitation energy and the conversion of zeaxanthin from violaxanthin in the light-harvesting complexes of plants [54–56]. Under these situations, violaxanthin is reversibly de-epoxidized by violaxanthin de-epoxidase to zeaxanthin [57–59]. To put it simply, violaxanthin becomes an efficient accessory pigment in weak light and zeaxanthin becomes an efficient photoprotector in strong light [60]. This association also has been relevant for a varied range of environmental conditions, for example, water stress and high temperature and not just under strong light exposure [61].

This chapter demonstrates that light exposure to Agria and Desiree minitubers leads to the similarity that both total and individual carotenoids were elevated up to two-fold higher on a μg/g DW basis than the total and individual carotenoids produced by dark treatment except for violaxanthin. The findings were consistent with the result reported in the previous studies [62, 63], where high violaxanthin was detected in sun-grown crop plants. However, they are not in agreement with Havaux and Niyogi [60] who found high violaxanthin in the dark and high zeaxanthin in strong light. Lutein and total carotenoid content also high in accordance with their observations and others [64–66]. In this study, lutein and total carotenoid in Agria and Desiree minitubers also increased with light. The changes were due to the stoichiometric and cyclical transformations among violaxanthin, antheraxanthin, and zeaxanthin [67]. Light encourages the de-epoxidase reaction and requires acidity for de-epoxidase activity, which can be caused by ATP hydrolysis or supplied by buffer [24, 68, 69]. The de-epoxidase is stereospecific for xanthophylls and as a consequence of that the carotenoid polyene chain must be all-trans. Otherwise, neoxanthin, which is 9-cis, is a passive substrate but becomes active when isomerized to the all-trans form [70].

The phytohormone abscisic acid (ABA) shows a regulatory role in most physiological processes in plants [71]. Various stress conditions, for instances, water, drought, cold, light, and temperature caused an increased amount of ABA. The action of ABA includes alteration of gene expression and analysis of responsive promoters discovered several potential *cis-* and *trans-*acting regulatory elements. In some of the controls in Agria and Desiree minituber experiments, zeaxanthin was detected. The occurrence of zeaxanthin might be in response to the brief exposure of samples to light. Every week, all minituber samples were checked and observed for contamination and size of minitubers. This brief exposure to light might trigger the accumulation of zeaxanthin in some of the minituber control samples.

The presence of zeaxanthin in Agria and Desiree minitubers developing on dark-grown plants could be justified by modification of gene expression in response to stress. Stress recognition may activate signal transduction pathways that transmit information inside the individual cell and through the plant. This may induce gene expression changes that influence growth and development as well as regulate the carotenoid biosynthesis. A stress will trigger and alter cellular metabolism, and as a result zeaxanthin accumulated as a precursor to ABA biosynthesis. Furthermore, plant resistance or sensitivity against stress can be determined by the species, genotype, and development age. In addition, there was a study revealing that different developmental age will accumulate different carotenoid [45]. In their study, 28 day stolons similar to our 4-week minitubers were detected to have highest total carotenoid compared to 80-day developing tubers and 9-month mature tubers. In both cases, zeaxanthin was also detected. Morris et al. [45] also demonstrated comparable results whereby the orange flesh tubers of DB375/1 were detected with high zeaxanthin, while pale yellow Desiree was detected with high violaxanthin. Besides, yellow flesh cultivars were observed to have the capability and ability to generate more carotenoids compared to the white flesh cultivars. Yellow flesh cultivars with high carotenoid content are able to tolerate stress, mainly light with tolerance mechanism [49]. As a result, Desiree minitubers accumulated violaxanthin and neoxanthin when stored in light, while Agria minitubers accumulated zeaxanthin and violaxanthin. In the experiment involving nutrient stress, the higher nutrient concentration given, the higher content of total and individual carotenoids of Agria found. On the other hand, in Desiree, total and individual carotenoids initially increased with increasing nutrient level, but then decreased at higher nutrient levels. The result showed that Desiree minitubers accumulated violaxanthin and neoxanthin, while Agria minitubers accumulated only violaxanthin.

Water deficiency is another significant environmental stress, which could influence plant growth and development as it is an essential element to meet basic needs. Drought, hypersaline environments, low temperatures, and transient loss of turgor at midday are among environmental conditions that can cause water deficit [49]. In a water stress condition, ABA will be synthesized by the roots and carried it into the shoots, with ABA being an important mediator to trigger plant responses, especially carotenoid biosynthesis to adverse environmental stimuli [72]. A major increase in the ABA content, particularly, in crops such as winter wheat, potatoes, and alfalfa was detected during hardening and cold acclimation [73–76]. However, the extent of the ABA response influenced by several differences [76], for instance, in winter wheat, a freeze resistant variety of wheat had a greater ABA level than a less resistant variety. Likewise, an increase in the ABA content was also observed in *Solanum commersonii,* but not in *S. tuberosum,* which failed to acclimate at −3°C. Besides, the total and individual carotenoid concentrations in both cultivars increased slightly during the PEG treatments, where the drought stress was simulated (**Figure 2**).

Oxidative cleavage is the first committed reaction for ABA biosynthesis in plant and it has commonly been assumed to be the key regulatory step. Various types of stress could encourage the ABA synthesis; therefore, ABA may be thought as a plant stress hormone [71]. ABA was known as one of the crucial plant growth and development regulator. A significant role for ABA in modulation at the gene level of adaptive responses of plants in adverse environmental conditions was also reported in several previous researches [77–80]. In some other physiological processes, ABA is also involved, for example, in leaf senescence [72], stomatal closure, embryo morphogenesis, development of seeds, and synthesis of storage proteins and lipids [81], germination [82], and defense against pathogens [83]. Nevertheless, ABA plays a role as a mediator in regulating adaptive plant responses to environmental stresses [84]. In certain cases, it has been involved in signal transduction at the single cell level [85]. Therefore, the findings of this study clearly demonstrated that environmental conditions for plant growth and development, such as light, dark, water stress, and nutrient concentration were significantly affecting and stimulating the carotenoid biosynthesis. At the same time, like other environmental stress response, disease or pathogen infection can also lead to oxidative stress responses, which implicate stress response genes [86], as well as storage period, which can cause physical changes such as sprouting and dehydration [87].

The results also suggested that ABA could facilitate a regulatory step for the carotenoid biosynthetic pathway versus environmental stress and during the first committed step in ABA biosynthesis, the epoxidation of zeaxanthin to violaxanthin by ZEP has happened. Zeaxanthin acts as a key element and indicator for the occurrence of environmental stress. In response to environmental stress conditions, violaxanthin and neoxanthin merely accumulated toward producing xanthoxin or precursors of ABA biosynthesis pathway. Predictably, the potato genotypes response to that environmental condition seemed to be highly genotype dependent and time duration exposed to stress. The activity of functional enzymes and candidate enzymes is another factor, which regulates carotenoid biosynthesis that determines the individual carotenoids type and quantity. Since the environmental conditions can influence carotenoid biosynthesis, undoubtedly, the carotenoid type and concentration that accumulates in potato tubers can be induced.

In a nutshell, the differences in carotenoid profile and tuber flesh color from different growing seasons, locations, and cultivars can be explained by the genes regulation, particularly ZEP and VDE, the existence of structure sequestering carotenoids, and the presence of the environmental stress. White flesh cultivars, which have a limited capacity to tolerate excessive light, exhibited an increased susceptibility to photooxidative damage [60]. In contrast, yellow flesh cultivars whose carotenoid content is much higher can specifically tolerate excessive light and also many environmental stress conditions by regulating ZEP and VDE. On top of better yield production, potatoes nutritional value and quality can be enhanced by selecting the appropriate potato cultivars that meet suitable environmental conditions for applicable agronomic practices. In simple words, to increase the accumulation of specific individual carotenoid pigments, implementing the most appropriate environmental factors is required. This invention could be more effective than selecting potato genotypes with higher carotenoid content as parents in a breeding program for the new potato cultivars development with enriched nutrients.

Acknowledgements

The authors are thankful to the Ministry of Education (MOE) and International Islamic University Malaysia (IIUM) for the Research Initiative Grant Scheme RIGS 16-396-0560.

Author details

Rashidi Othman[1]*, Norazian Mohd Hassan[2] and Farah Ayuni Mohd Hatta[1]

*Address all correspondence to: rashidi@iium.edu.my

1 International Institute for Halal Research and Training (INHART), Herbarium Unit, Department of Landscape Architecture, Kulliyyah of Architecture and Environmental Design, International Islamic University Malaysia, Kuala Lumpur, Malaysia

2 Department of Pharmaceutical Chemistry, Kulliyyah of Pharmacy, International Islamic University Malaysia, Kuala Lumpur, Malaysia

References

[1] Lu W, Haynes K, Wiley E, Clevidence B. Carotenoid content and colour in diploid potatoes. J. Amer. Soc. Hort. Sci. 2001; 126: 722–726.

[2] Brown CR, Edwards CG, Yang C-P, Dean BB. Orange flesh trait in potato: Inheritance and carotenoid content. J. Amer. Soc. Hort. Sci. 1993; 118:145–150.

[3] Römer S, Lübeck J, Kauder F, Steiger S, Adomat C, Sandmann G. Genetic engineering of a zeaxanthin-rich potato by antisense inactivation and co-suppression of carotenoid epoxidation. Metab. Eng. 2002; 4(4):263–272.

[4] Ducreux LJM, Morris WL, Hedley PE, Shepherd T, Davies HV, Millam S, Taylor MA. Metabolic engineering of high carotenoid potato tubers containing enhanced levels of β-carotene and lutein. J. Exp. Bot. 2004; 409:81–89.

[5] Hsi DCH, Young CT, Ortiz M. Effect of growing seasons, locations and planting dates on total amino acid composition of two Valencia peanut varieties grown in New Mexico. Peanut Sci. 1981; 8:131.

[6] Oupadissakoon C, Young CT, Geisbrecht FG, Perry A. Effect of location and time of harvest on free amino acid and free sugar contents of Florigiant peanuts. Peanut Sci. 1980; 7:61.

[7] Dawson R, McIntosh AD. Varietal and environmental differences in the proteins of the groundnut (*Arachis hypogaea* L.). J. Sci. Food Agric. 1973; 24:597.

[8] Amaya-F J, Young CT, Mixon AC, Nordon AJ. Soluble amino and carbohydrate compounds in the testae of six experimental peanut lines with various degrees of Aspergillus flauus resistance. J. Agric. Food Chem. 1977; 25: 661.

[9] Amaya-F J, Basha SMM, Young CT. Variation in total monosaccharides in developing peanuts (*Arachis hypogaea* L). Cienc. Cult. 1978; 30:79.

[10] Basha SMM, Cherry JP, Young CT. Changes in free amino acids, carbohydrates, and proteins of maturing seeds from various peanut (*Arachis hypogaea* L.) cultivars. Cereal Chem. 1976; 53: 586.

[11] Young CT, Waller GR, Matlock RS, Morrison RD, Hammons RO. Some environmental factors affecting free amino acid composition in six varieties of peanuts. J. Am. Oil Chem. Soc. 1974; 51:265.

[12] Young CT, Matlock RS, Mason ME, Waller GR. Effect of harvest date and maturity upon free amino acid levels in three varieties of peanuts. J. Am. Oil Chem. Soc. 1974; 51:269.

[13] Young CT. Amino acid composition of peanut (*Arachis hypogaea* L.) samples from the 1973 and 1974 uniform peanut performance tests. Proc. Am. Peanut Res. Educ. Soc. 1979; 11: 24.

[14] Friedman M, McDonald GM. Potato glycoalkaloids: Chemistry, analysis, safety and plant physiology. Crit. Rev. Plant Sci. 1997; 16: 55–132.

[15] Sanford LL, Sinden SL. Inheritance of potato glycoalkaloids. Am. Potato J. 1972; 49:209–217.

[16] Sanford LL, Deahl KL, Sinden SL, Kobayashi RS. Glycoalkaloid content in tubers of a hybrid and backcross populations from *Solanum tuberosum* (×) *chaconense* cross. Am. Potato J. 1995; 72:261–271.

[17] Haynes KG, Sieczka JB, Henninger MR, Fleck DL. Clone × environment interactions for yellow-flesh intensity in tetraploid potatoes. J. Amer. Soc. Hort. Sci. 1996; 121(2): 175–177.

[18] Griffiths DW, Dale MFB, Morris WL, Ramsay G. Effects of season and postharvest storage on the carotenoid content of Solanum phureja potato tubers. J. Agric. Food Chem. 2007; 55: 379–385.

[19] Anderson KA, Smith BW. Effect of season and variety on the differentiation of geographic growing origin of pistachios by stable isotope profiling. J. Agric. Food Chem. 2006; 54: 1747–1752.

[20] Jing PU, Noriega V, Schwartz SJ, Giusti MM. Effects of growing conditions on purple corncob (*Zea mays* L.) anthocyanins. J. Agric. Food Chem. 2007; 55: 8625–8629.

[21] Fraser PD, Bramley PM. The biosynthesis and nutritional uses of carotenoids. Prog. Lipid Res. 2004; 43: 228–265.

[22] Chloupek O, Hrstkova P. Adaptation of crops to environment. Theor. Appl. Genet. 2005; 111: 1316–1321.

[23] Othman R. Biochemistry and genetics of carotenoid composition in potato tubers [thesis]. Christchurch, New Zealand: Lincoln University; 2009.

[24] Rockholm DC, Yamamoto HY. Violaxanthin de-epoxidase. Purification of a 43-kilodalton lumenal protein from lettuce by lipid-affinity precipitation with monogalactosyldiacylglyceride. Plant Physiol. 1996; 110: 697–703.

[25] Cunningham FX Jr, Gantt E. Genes and enzymes of carotenoid biosynthesis in plants. Ann. Rev. Plant Physiol. Plant Mol. Biol. 1998; 49: 557–583.

[26] Howitt CA, Pogson BJ. Carotenoid accumulation and function in seeds and non-green tissues. Plant Cell Environ. 2006; 29:435–445.

[27] Hager A Die reversiblen, lightabhangigen Xanthophyllumwandlungen im Chloroplasten. Ber Dtsch Bot Ges. 1975; 88: 2744.

[28] Siefermann D, Yamamoto HY. Properties of NADPH and oxygen-dependent zeaxanthin epoxidation in isolated chloroplasts. A transmembrane model for the violaxanthin cycle. Arch. Biochem. Biophys. 1975; 171(1): 70–77.

[29] Ruban AV, Young AJ, Pascal AA, Horton P. The effects of illumination on the xanthophyll composition of the photosystem II light-harvesting complex of spinach thylakoid membranes. Plant Physiol. 1994; 104: 227–234.

[30] Lee AI-C, Thornber JP. Analysis of the pigment stoichiometry of pigment-protein complexes from barley (*Hordeum vulgare*). Plant Physiol. 1995; 107:565–574.

[31] Verhoeven AS, Adams III WW, Demmig-Adams B, Croce R, Bassi R. Xanthophyll cycle pigment localization and dynamics during exposure to low temperatures and light stress in vinca major. Plant Physiol. 1999; 120:727–737.

[32] Yamamoto HY. Biochemistry of the violaxanthin cycle in higher plants. Pure Appl. Chem. 1979; 51: 639–648.

[33] Yamamoto HY . Xanthophyll cycles. Methods Enzymol. 1985;110: 303–312.

[34] Croce R, Weiss S, Bassi R. Carotenoid-binding sites of the major light-harvesting complex II of higher plants. J. Biol. Chem. 1999; 274: 29613–29623.

[35] Hobe S, Niemeier H, Bender A, Paulsen H. Carotenoid binding sites in LHCIIb: Relative affinities towards major xanthophylls of higher plants. Eur. J. Biochem. 2000; 267: 616–624.

[36] Pogson B, McDonald KA, Truong M, Britton G, DellaPenna D. Arabidopsis carotenoid mutants demonstrate that lutein is not essential for photosynthesis in higher plants. Plant Cell. 1996; 8: 1627–1639.

[37] Pogson B, Niyogi KK, Björkman O, DellaPenna D. Altered xanthophyll compositions adversely affect chlorophyll accumulation and nonphotochemical quenching in Arabidopsis mutants. Proc. Natl. Acad. Sci. U. S. A. 1998; 95: 13324–13329.

[38] Tardy F, Havaux M. Photosynthesis, chlorophyll fluorescence, light-harvesting system and photoinhibition resistance of a zeaxanthin-accumulating mutant of *Arabidopsis thaliana*. J. Photochem. Photobiol. 1996; 34: 87–94.

[39] Heinze I, Pfündel E, Hühn M, Dau H. Assembly of light harvesting complexes II (LHC-II) in the absence of lutein. A study on the α-carotenoid-free mutant C-2A'-34 of the green alga Scenedesmus obliquus. Biochim. Biophys. Acta. 1997; 1320: 188–194.

[40] Farber A, Young AJ, Ruban AV, Horton P, Jahns P. Dynamics of xanthophyll-cycle activity in different antenna subcomplexes in the photosynthetic membranes of higher plants. The relationship between zeaxanthin conversion and nonphotochemical fluorescence quenching. Plant Physiol. 1997; 115: 1609–1618.

[41] Polle JEW, Niyogi KK, Melis A. Absence of lutein, violaxanthin and neoxanthin affects the functional chlorophyll antenna size of photosystem-II but not that of photosystem-I in the green alga *Chlamydomonas reinhardtii*. Plant Cell Physiol. 2001; 42: 482–491.

[42] Morosinotto T, Caffarri S, Dall'Osto L, Bassi R. Mechanistic aspects of the xanthophyll dynamics in higher plant thylakoids. Physiol. Plant. 2003; 119: 347–354.

[43] Murashige T, Skoog F. A revised medium for rapid growth and bioassays with tobacco tissue cultures. Physiol. Plant. 1962; 15:473–497.

[44] Conner AJ, Williams MK, Gardner RC, Deroles SC, Shaw ML, Lancaster JE. Agrobacterium-mediated transformation of New Zealand potato cultivars. N. Z. J. Crop Hortic. Sci. 1991 ;19:1–8.

[45] Morris WL, Ducreux L, Griffiths DW, Stewart D, Davies HV, Taylor MA. Carotenogenesis during tuber development and storage in potato. J. Exp. Bot. 2004; 55: 975–982.

[46] Lewis DH, Bloor SJ, Schwinn KE. Flavonoid and carotenoid pigments in flower tissue of *Sandersonia aurantiaca* (Hook.). Sci. Hortic. 1998; 72:179–192.

[47] Britton G, Structure and properties of carotenoids in relation to function. FASEB J. 1995; 9: 1551–1558.

[48] Kimura M, Rodriguez-Amaya DB. A scheme for obtaining standards and HPLC quantification of leafy vegetable carotenoids. Food Chem. 2002; 78:389–398.

[49] Buchanan BB, Gruissem W, Jones RL. Biochemistry and Molecular Biology of Plants. Rockville, MD: American Society of Plant Biologists; 2000.

[50] Siefermann-Harms D. The light harvesting and protective functions of carotenoids in photosynthetic membranes. Physiol. Plant. 1987; 69:561–568.

[51] Frank HA, Cogdell RJ. Carotenoids in photosynthesis. Photochem. Photobiol. 1996; 63:257–264.

[52] Telfer A, Dhami S, Bishop SM, Phillips D, Barber J. β-Carotene quenches singlet oxygen formed by isolated photosystem II reaction centers. Biochemistry. 1994; 33:14469–14474.

[53] Bramley PM, Mackenzie A. Regulation of carotenoid biosynthesis. Curr. Topics Cell. Regn. 1987; 29:291.

[54] Young AJ, Phillip, D, Ruban AV, Horton P, Frank HA. The xanthophyll cycle and carotenoid-mediated dissipation of excess excitation energy in photosynthesis. Pure Appl. Chem. 1997; 69(10):2125–2130.

[55] Baker NR, Bowyer JR. Photoinhibition of Photosynthesis. From Molecular Mechanisms to the Field. Oxford, UK: Bios Scientific Publishers; 1994.

[56] Long SP, Humphries S, Falkowski PG. Photoinhibition of photosynthesis in nature. Annu. Rev. Plant Physiol. Plant Mol. Biol. 1994; 45:633–662.

[57] Pfundel E, Bilger W. Regulation and possible function of the violaxanthin cycle. Photosynth. Res. 1994; 42:89–109.

[58] Demmig-Adams B, Adams W III. The role of xanthophyll cycle carotenoids in the protection of photosynthesis. Trends Plant Sci. 1996; 1:21–26.

[59] Eskling M, Arvidsson PO, Akerlund HE. The xanthophyll cycle, its regulation and components. Physiol. Plant. Physiol. Plant. 1997; 100: 06–816.

[60] Havaux M, Niyogi KK. The violaxanthin cycle protects plants from photooxidative damage by more than one mechanism. Proc. Natl. Acad. Sci. U.S.A. 1999; 96:8762–8767.

[61] Demmig-Adams B, Adams WWIII. The xanthophyll cycle, protein turnover, and the high light tolerance of sun-acclimated leaves. Plant Physiol. 1993; 103(4):1413–1420.

[62] Demmig-Adams B, Adams WWIII. Carotenoid composition in sun and shade leaves of plants with different life forms. Plant Cell Environ. 1992; 15:411–419.

[63] Sapozhnikov DI, Krasovskaya TA, Maevskaya AN. Change in the interrelationship of the basic carotenoids of the plastids of green leaves under the action of light. Dokl. Akad. Nauk. USSR. 1957; 113:465–467.

[64] Thayer SS, Bjorkman O. Leaf xanthophyll content and composition in sun and shade leaves determined by HPLC. Photosynth. Res. 1990; 23:331–343.

[65] Demmig-Adams B, Adams WWIII. Photoprotection and other responses of plants to high light stress. Annu. Rev. Plant Physiol. Plant Mol. Biol. 1992; 43:599–626.

[66] Johnson GN, Scholes JD, Horton P, Young AJ. Relationships between carotenoid composition and growth habit in British plant species. Plant Cell Environ. 1993; 16:681–686.

[67] Yamamoto HY, Nakayama TOM, Chichester CO. Studies on the light and dark interconversions of leaf xanthophylls. Arch. Biochem. Biophys. 1962; 97:168–173.

[68] Hager A. Lichtbedingte pH-Erniedringung in einem Chloroplasten-Kompartiment als Ursache der enzymatischen Violaxanthin + Zeaxanthin-Umwandlung; Beziehungen zur Photophosphorylierung. Planta. 1969; 89:224–243.

[69] Yamamoto HY, Kamite L, Wang YY. An ascorbate-induced absorbance change in chloroplasts from violaxanthin de-epoxidation. Plant Physiol. 1972; 49:224–228.

[70] Yamamoto HY, Higashi RM. Violaxanthin de-epoxidase: Lipid composition and substrate specificity. Arch. Biochem. Biophys. 1978; 190:514–522.

[71] Swamy PM, Smith BN. Role of abscisic acid in plant stress tolerance. Curr. Sci. 1999; 76:1220–1227.

[72] Zeevaart JAD, Creelman RA. Metabolism and physiology of abscisic acid. Annu Rev. Plant Physiol. Plant Mol. Biol. 1988; 39:439–473.

[73] Chen THH, Li PH, Brenner ML. Involvement of abscisic acid in potato cold acclimation. Plant Physiol. 1983; 71:362–365.

[74] Luo M, Liu JH, Mahapatra S, Hiu RD, Mahapatra SS. Characterization of a gene family encoding abscisic acid- and environmental stress-inducible proteins of alfafa. J. Biol. Chem. 1992; 267:432–436.

[75] Wrightman F. Plant Regulation and World Agriculture. New York: Plenum Press. 1979; pp. 324–377.

[76] Lalk I, Dorffling K. Hardening, abscisic acid, praline and freezing resistance in two winter wheat varieties. Physiol. Plant. 1985; 63:287–292.

[77] Orr W, Keller WA, Singh J. Induction of freezing tolerance in an embryonic cell suspension culture of *Brassica napus* by abscisic acid at room temperature. Plant Physiol. 1986; 126:23–32.

[78] Ramagopal S. Differential mRNA transcription during salinity stress in barley. Proc. Natl. Acad. Sci. U. S. A. 1987; 84:94–98.

[79] Singh NK, Bracker CA, Hasigawa PM, Bressan RA. Characterization of osmotin: a thaumatin-like protein associated with osmotic adaptation in plant cells. Plant Physiol. 1987; 85:529–536.

[80] Pena-Cortes H, Sanchez-Serrano J, Mertens R, Willmitzer L, Prat S. Abscisic acid is involved in the wound-induced expression of the proteinase inhibitor II gene in potato and tomato. Proc. Natl. Acad. Sci. U. S. A. 1989; 86:9851–9855.

[81] Thomas TL. Gene expression during plant embryogenesis and germination: an overview. Plant Cell. 1993; 5:1401–1410.

[82] Koornneef M, Hanhart CJ, Hirost HWM, Karssen CM. In vivo inhibition of seed development and reserve protein accumulation in recombinants of abscisic acid biosynthesis and responsiveness mutants in *Arabidopsis thaliana*. Plant Physiol. 1989; 90:462–469.

[83] Dunn RM, Hedden P, Bailey JA. A physiologically-induced resistance of Phaseolus vulgaris to a compatible race of Colletotrichum lindemuthianum is associated with increases in ABA content. Physiol. Mol. Plant Pathol. 1990; 36:339–349.

[84] Ingram J, Bartel D. The molecular basis of dehydration tolerance in plants. Annu. Rev. Plant Physiol. Plant Mol. Biol. 1996; 47:377–403.

[85] Jeffrey L, Giraudat J. Abscisic acid signal transduction. Annu. Rev. Plant Physiol. Plant Mol. Biol. 1998; 49:199–222.

[86] Desender S, Andrivon D, Val F. Activation of defence reactions in Solanaceae: Where is the specificity? Cell. Microbiol. 2007; 9:21–30.

[87] Blessington T, Miller Jr JC, Nzaramba MN, Hale AL, Redivari L, Scheuring DC, Hallman GJ. The effects of low dose gamma irradiation and storage time on carotenoids, antioxidant activity, and phenolics in the potato cultivar Atlantic. Amer. J. Potato Res. 2007; 84:125–131.

Chapter 7

Carotenoids Regulate Endothelial Functions and Reduce the Risk of Cardiovascular Disease

Kazuo Yamagata

Additional information is available at the end of the chapter

http://dx.doi.org/10.5772/67464

Abstract

Regular consumption of fruits and vegetables can help reduce the risk for cardiovascular disease (CVD) and its associated mortality. A diet rich in fruits and vegetables is thought to have cardioprotective effects, but the specific components of these foods that provide this protection are unclear. Antioxidants such as vitamin C, carotenoids, and polyphenols in fruits and vegetables likely contribute to the reduction in risk of CVD by minimizing cholesterol oxidation in blood vessel walls. Meanwhile, cardioprotective effects afforded by the carotenoids lycopene, α-carotene, β-carotene, β-cryptoxanthin, lutein, and zeaxanthin have been reported in many studies. Carotenoids are naturally occurring fat-soluble pigments that are present at high levels in tomatoes and carrots. Carotenoids play an important role in staving off atherosclerosis via antioxidant activities that reduce lipid peroxidation in low-density lipoproteins. Lycopene reduces endothelin-1 gene expression by suppressing generation of reactive oxygen species and inducing heme oxygenase-1 expression in human endothelial cells. Thus, carotenoids may mitigate endothelial dysfunction by promoting direct antioxidative effects and inducing expression of several genes. Structural and functional differences among carotenoids may explain their unique biologic activities. In this review, the roles of carotenoids in relation to their influence on vascular endothelial functions and cardioprotective effects are discussed.

Keywords: carotenoids, cardiovascular disease, endothelial cells

1. Introduction

Cardiovascular disease (CVD) is a common disease that has high mortality. Many epidemiological studies indicate that a diet rich in fruits and vegetables can have preventive effects for the development of CVD [1, 2]. As such, sufficient consumption of fruits and vegetables is recommended to ensure that vitamins, fiber, potassium, folate, and phenolic molecules are present in proper amounts to yield health benefits [3]. Several of these nutritive components have antioxidant activity and can modify lipoprotein profiles as well as increase insulin sensitivity, and lower blood pressure [4, 5]. Although carotenoids in particular are thought to provide health benefits, several studies suggested that these preventative effects may not be due to β-carotene and vitamin E present in fruits and vegetables [6]. In fact, some reports demonstrated that other carotenoids such as lycopene in tomatoes have preventive effects for CVD [7, 8].

Dietary carotenoids primarily come from fruits and vegetables, as well as plant seeds, roots, leaves, and flowers. Among 12 types of dietary carotenoids, particularly α-carotene, β-carotene, lycopene, lutein, β-cryptoxanthin, and zeaxanthin, can be found in human blood and tissue samples [9, 10], and these molecules have similar chemical constitutions (**Figure 1**) and health benefits [11] (**Table 3**). α-Carotene, β-carotene, γ-carotene, lycopene, and β-cryptoxanthin are all precursors of vitamin A. These carotenoids also have other beneficial effects beyond their antioxidant activity [12, 13].

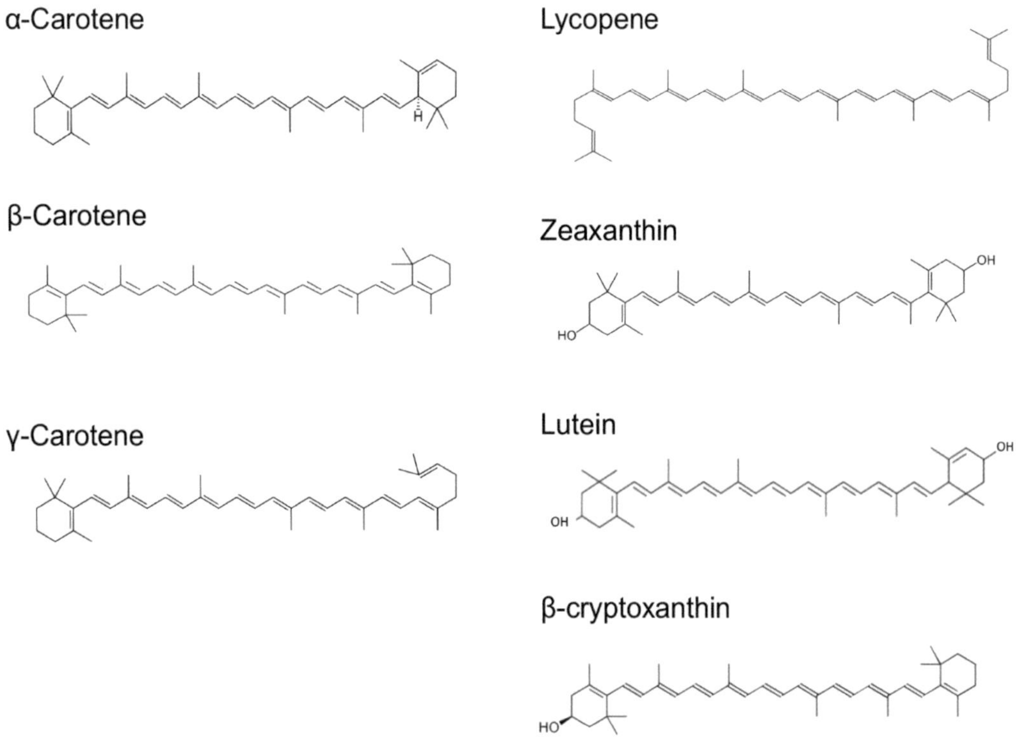

Figure 1. Chemical structures of several carotenoids.

Vascular endothelial cell disorders are a hallmark CVD. Several epidemiologic studies indicate that carotenoids can have a beneficial effect on vascular endothelial cell dysfunction. For example, in experiments using cultured vascular endothelial cells, carotenoids regulated nitric oxide (NO) expression and endothelin-1 (ET-1) production [14]. Moreover, lycopene inhibits expression of lipopolysaccharide (LPS)-enhanced monocyte chemoattractant protein-1 (MCP-1), interleukin-6 (IL-6), and vascular cell adhesion molecule-1 (VCAM-1) in human endothelial cells [14].

In contrast, lycopene reduced expression of TNF-α–induced intercellular adhesion molecule-1 (ICAM-1) and adhesion of monocyte endothelial cells [15]. In streptozotocin (STZ)-induced diabetic rats, lycopene inhibited endothelial dysfunction [16]. However, the *in vitro* effects of dietary carotenoids do not always translate to an *in vivo* setting. In the present review, we discuss the influence of carotenoids on vascular endothelial functions. Furthermore, we summarize evidence that carotenoids may have a preventive benefit toward CVD.

2. Source and bioactivity of natural carotenoids

Carotenoids are found as α-carotene, β-carotene, lycopene, lutein, β-cryptoxanthin, and zeaxanthin. Carotenoids are tetraterpenoids and are synthesized in plants such as vegetables and fruits as well as by other photosynthetic organisms and some nonphotosynthetic bacteria, yeasts, and molds [17]. Carotenoids confer the orange, yellow, and red color of many fruits and vegetables. Carotenoids can be classified as carotenes and xanthophylls according to the chemical structure. Xanthophylls contain oxygen, whereas carotenes are purely hydrocarbons and lack oxygen. The structures of common carotenoids are shown in **Figure 1**. β-Carotene is the most commonly found carotenoid in raw vegetables, canned fruits, and cooked vegetables [13]. Lycopene is present in tomato-based foods, including tomato paste, catsup, and other processed tomato products. Zeaxanthin and lutein are found in cooked kale and spinach and in a number of processed spinach products. Carotenoids can also be found in insects, fish, and crustaceans. The main sources and contents of dietary carotenoids are listed in **Tables 1** and **2** [13, 18]. Carotenoids can be classified into pro-vitamin A and nonpro-vitamin A groups [19]. Daily vitamin A intake is dependent on the pro-vitamin A content of foods. In developing countries, approximately 70% of vitamin A intake is derived from carotenoids found in vegetables and fruits [17]. Pro-vitamin A is converted into vitamin A in the body via mechanisms that are not fully characterized, such that for purposes of bioequivalence, vitamin A levels are quantified according to vitamin A intake. Moreover, conversion efficiencies from carotenoid to vitamin A may influence the biological activity of carotenoids [20].

Carotenoid	Source
β-Carotene	Carrots, apricots, mangoes, red pepper, kale, spinach, broccoli
α-Carotene	Carrots, collard greens, pumpkin, corn, yellow pepper
β-Cryptoxanthin	Avocado, oranges, papaya, passion fruit, pepper, persimmon
Lutein plus zeaxanthin	Kale, spinach broccoli, peas, brussels sprouts, collard greens, lettuce, corn, egg yolk
Lycopene	Tomato and tomato products, watermelon, pink grapefruit, papaya, guava, rose hip

Voutilainen et al. [17].

Table 1. Sources of dietary carotenoids.

Carotenoids	Food	Content (mg/100 g wet wt)[a]
β-Carotene	Carrots, raw	18.3
	Mangos, canned	13.1
	Sweet potato, cooked	9.5
	Carrots, cooked	8.0
	Pumpkin, canned	6.9
	Kale, cooked	6.2
	Spinach, cooked	5.2
	Winter butternut squash	4.6
	Swiss chard, raw	3.9
	Apricots, raw	2.6
	Pepper, red, raw	2.4
	Pepper, red, cooked	2.2
	Cantaloupe, raw	1.6
	Lettuce, romaine, raw	1.3
	Tomato paste	1.2
Lycopene	Tomato paste	29,3
	Catsup	17.0
	Tomato puree	16.7
	Pasta sauce	16.0
	Tomato sauce	15.9
	Tomato soup	10.9
	Tomato, canned, whole	9.7
	Tomato juice	9.3
	Watermelon, raw	4.9
	Tomato, cooked	4.4
	Tomato, raw	3.0

Carotenoids	Food	Content (mg/100 g wet wt)[a]
Lutein and zeaxanthin	Kale, cooked	15.8
	Spinach, raw	11.9
	Spinach, cooked	7.0
	Lettuce, romaine, raw	2.6
	Broccoli, raw	2.4
	Broccoli, cooked	2.2
	Summer squash, zucchini	2.1
	Corn, sweet, cooked	1.8
	Peas, green, canned	1.4
	Brussels sprouts, cooked	1.3
	Corn, sweet, canned	0.9
	Beans, green, cooked	0.7
	Beans, green, canned	0.7
	Beans, green, raw	0.6
	Okra, cooked	0.4
	Cabbage, white, raw	0.3
	Egg yolk, medium	0.3
	Celery, raw	0.2
	Orange, raw	0.2
	Tomato paste	0.2

[a]Edible portion.
Krinsky and Johnson [13].

Table 2. Carotenoid contents in foods.

3. Epidemiological studies of carotenoids

Many epidemiologic studies showed that carotenoids have beneficial effects toward CVD (**Table 3**). A cohort study that included 91,379 men, 129,701 women, and 5007 coronary heart disease events showed that fruits and vegetables intake was associated with decreased levels of coronary heart disease [2]. Meanwhile, another large cohort study indicated that fruits and vegetables intake can reverse coronary heart disease [21]. Many epidemiological studies indicated that higher serum carotenoid levels have beneficial effects on CVD biomarkers. For example, lycopene intake was associated with decreased levels of CVD in a study of 314 CVD patients, 171 CHD patients, and 99 stroke patients [22]. Hazard ratios (HRs) for CVD onset were inversely correlated with lycopene intake. Another study that examined the intake of dietary carotene by 1312 men and 1544 women showed that dietary lutein and zeaxanthin consumption was clearly related to CVD onset, risk ratios, and biomarker levels such as HDL cholesterol [23]. A significant inverse relationship between LDL cholesterol and

Intake from dietary	Study name	Nationality of subjects	Follow-up, Time	The number of subjects	Sex	Outcome (main results)	Reference (author, issue year)
Carotenoids with provitamin A activity	Finnish Mobile Clinic Study	Finnish	Prospective, 14 y	5133	F, M	Coronary mortality (nonsignificant inverse association between dietary intake of carotenoids with provitamin A activity and the risk of coronary mortality in women)	Knekt et al., 1994
Carotenoids with provitamin A activity	ARIC study	American	Cross-sectional	12,773	F, M	Prevalence of carotid plaques (those in the highest quintile of carotenoid consumption had a lower prevalence of plaques)	Kritchevsky et al., 1998
β-Carotene	The Rotterdan study	Dutch	Prospective, 4 y	4802	F, M	Myocardial infarction (significantly decreased risk of myocardial infarction in highest -carotene intake quartile)	Klipstein-Grobusch et al., 1999
β-Carotene, lutein plus zeaxanthin, and lycopene	ATBC study	Finnish	Prospective, 6.1 y	26,593	M	Stroke (dietary intake of β-carotene was inversely associated with the risk of cerebral infarction)	Hirvonen et al., 2000
α- and β-Carotene, lutein αplus zeaxanthin, lycopene, and cryptoxanthin	Nurses Health Study	American	Prospective, 12 y	73,286	F	Coronary artery disease (inverse significant associations between the highest quintiles of intake of α-carotene and β-carotene and risk of coronary artery disease)	Osganian et al., 2003

Voutilainen et al. [17].

Table 3. Epidemiological studies of the effect on cardiovascular disease and atherosclerosis with carotenoids.

β-carotene, lutein, and zeaxanthin consumption as well as levels of dietary β-carotene and homocysteine was observed, whereas serum β-carotene affected the relationship between dietary β-carotene intake and C-reactive protein (CRP) levels. Given that hyperlipidemia, serum CRP, and homocysteine are CVD onset risk factors, serum carotenoids may be markers of dietary carotenoid uptake and CVD risk biomarkers. Indeed, a report by Sesso et al. [7] found that higher plasma lycopene levels were associated with decreased risk of CVD in a survey of 39,876 elderly women. In addition, a prospective study indicated that plasma α-carotene, β-carotene, and lycopene levels were associated with the risk of ischemic stroke [24]. A population-based follow-up study in Japan that examined the relationship between CVD and carotene concentration in 3061 subjects showed that higher serum total carotene levels, including α- and β-carotene and lycopene levels were linked with a reduced risk of CVD mortality [8]. Furthermore, report the inverse significant associations between the highest quintiles of the intake of α-carotene and β-carotene and risk of coronary artery disease [25]. In addition, dietary intake of β-carotene was inversely associated with the risk of cerebral infarction [26].

Marine animals produce the carotenoid astaxanthin that is known to have strong antioxidative activity. A study of 24 volunteers that consumed increasing doses of astaxanthin over the course of 14 days showed inhibition of LDL oxidation relative to control subjects that did not consume astaxanthin [27].

In contrast, other reports indicated that fruits and vegetables consumption is not associated with a reduced risk of coronary heart disease [28]. In a study of overweight adults at high risk for CVD, no dose-dependent reduction in CVD risk factors was seen with increased fruits and vegetables intake [29]. These results indicate that there may be some restrictions in the degree of protection afforded by carotenoids [30]. Moreover, a study of healthy adult subjects showed no effects of lutein, lycopene, or β-carotene on biological markers of oxidative stress, including LDL oxidation [31]. In a prospective study, the relationship between plasma lycopene concentration and CVD risk in 499 men showed that higher plasma lutein, zeaxanthin, and retinol levels were associated with a moderate increase in CVD risk, whereas β-cryptoxanthin, α-carotene, and β-carotene were not associated with increased risk of CVD [32]. Likewise, a prospective study involving a population of male physicians in the United States showed that high plasma levels of retinol and carotenoids had no protective effect toward myocardial infarction [33]. Moreover, four extensive, randomized studies revealed no decrease in CVD events by β-carotene treatment [34, 35]. These conflicting results again suggest that the reduction in the risk of CVD associated with fruits and vegetables intake is so far largely confined to observational epidemiology [30].

4. Protective effects of carotenoid-enriched foods

4.1. Tomato carotenoids

Tomato intake has been hypothesized to prevent endothelial dysfunction. However, one study involving 19 postmenopausal women who ingested tomato puree had increased plasma lycopene levels, but no changes in artery dilation, which suggested that lycopene may not have direct effects on endothelial function [36]. On the other hand, another report demonstrated that tomato extract enhanced nitric oxide (NO) production and decreased endothelin release. These effects of tomato extract were related to suppression of inflammatory NF-κB signaling and prevention of adhesion molecule expression in endothelial cells [37], whereas tomato paste supplementation modified endothelial dysfunction and affected oxidation markers in the plasma of healthy human volunteers enrolled in a recent study [38]. Thus, these studies indicated that tomato paste intake can induce beneficial outcomes on endothelial function. The antioxidant properties of lycopene and β-carotene in tomato products may indeed regulate endothelial functions and protect against CVD. In a study that examined pigs with high cholesterol levels, consumption of a tomato-derived lycopene supplement maintained endothelial function of coronary arteries and regulated expression of apolipoprotein A-I and apolipoprotein J [39]. Lycopene supplementation also prevented vasoactive drug-induced coronary vasodilation and reduced lipid peroxidation, while enhancing high-density lipoprotein (HDL) levels and endothelial nitric oxide synthase (eNOS) expression. These results demonstrate that lycopene supplementation likely can protect against LDL-enhanced coronary endothelial dysfunction by augmenting endothelial nitric oxide (NO) expression and HDL levels as well as mediating leukocyte adhesion to endothelial cells in response to inflammation.

4.2. Carrot carotenoids

Carotenoids contained in carrots have beneficial health effects [40]. For example, drinking carrot juice induces antioxidant activity and reduces lipid peroxidation and can decrease levels of CVD risk markers in adults. In addition, carrot juice intake reduces systolic blood pressure [41]. Carrot juice consumption also improved glucose tolerance and hepatic structure and function, which might be associated with the effect of anthocyanins seen in metabolic syndrome [40, 42].

5. Preventive effects of carotenoids on cardiovascular disease associated with endothelial cell and macrophage dysfunction

Tomato paste supplementation regulated endothelial cell functions and prevented oxidative conditions in 19 healthy subjects [38]. Enhanced reactive oxygen species (ROS) generation is related to a functional inactivation of NO in endothelial cells and can induce CVD. β-Carotene and lycopene-mediated prevention of TNF-α expression was associated with reduced nitro-oxidative stress and inflammatory response in endothelial cells [43]. Meanwhile, in human endothelial cells, lycopene prevents endothelin-1 expression by inhibiting ROS generation and inducing heme oxygenase-1 expression (HO-1) [44], while also inhibiting tumor necrosis factor (TNF)-α–induced NF-κB activation, ICAM-1 expression, and monocyte endothelial adhesion [15]. In an *in vivo* study, lycopene inhibited endothelial dysfunction in STZ-enhanced diabetic rats by lowering oxidative stress, which could have implications for the development of treatments to prevent diabetic vascular complications [16]. In addition, astaxanthin inhibits inflammation-induced inducible NO and ROS generation by suppressing NF-κB pathway activity in macrophages [45]. Thus, carotenoids could be effective for treating diseases associated with oxidative stress, such as CVD [46].

In vitro studies indicated that endothelial dysfunction induces atherogenic risk [47]. As shown in **Table 4**, carotenoids have a beneficial effect on endothelial cell function. In a study of healthy men, lycopene supplementation was suggested to inhibit oxidative stress-mediated decreases in endothelium function [48]. For example, lycopene prevents LPS-induced MCP-1, IL-6, and VCAM-1 expression in human endothelial cells [14]. Similarly, lycopene inhibits activity of an LPS-enhanced proinflammatory cytokine cascade in human endothelial cells through a mechanism that may involve increased expression of Krüppel-like factor 2 (KLF2) and inhibition of toll-like receptor (TLR) 4 function as well as downstream extracellular signal-regulated kinase (ERK) and NF-κB signaling in human endothelial cells [14].

As mentioned above, ET-1 is a strong vasopressor produced by endothelial cells. ET-1 levels may be affected by lycopene and in turn reduce the risk of CVD by modulating the activity of antiinflammatory pathways. Indeed, one report indicated that lycopene prevents cyclic strain-induced endothelin-1 expression by suppressing ROS production in human endothelial cells [44]. Furthermore, β-carotene and lycopene reduced TNF-α–enhanced inflammatory responses by reducing nitro-oxidative stress. These functions decreased interactions of endothelial cells with monocytes [43]. Another report demonstrated that β-carotene and lycopene treatment reduced TNF-α–induced oxidative stress and inflammatory responses to affect interactions between monocytes and human endothelial cells [43]. Furthermore, lycopene reduces C-reactive protein levels in CVD [49]. Meanwhile, paraoxonase-1 (PON1) prevents the oxidation of lipoproteins induced by oxidative stress and may induce metabolism of lipid peroxides [50]. We demonstrated that β-carotene decreases IL-1β–induced downregulation in PON1 expression by activating the CaMKKII signaling pathway in human endothelial cells that may in turn produce antioxidant activity [51]. Similarly, astaxanthin reduces ROS induced-associated dysfunction in human endothelial cells exposed to glucose [52]. Astaxanthin inhibits streptozotocin-induced endothelial dysfunction in diabetes in male rats [53]. Astaxanthin also has antioxidant activity in human endothelial cells that is related to induction of p22phox expression and reduced peroxisome proliferator activated receptor-γ coactivator (PGC-1α) expression [54]. Together these activities of carotenoids may be responsible for their protective effect on CVD risk.

In cultured mouse macrophages, lutein-induced matrix metalloproteinase (MMP)-9 expression and phagocytosis promoted by intracellular ROS and activation of ERK1/2, p38 MAPK, and RAR β [55]. Furthermore, carotenoids induce increases in intracellular glutathione levels by elevating the activity of glutamate–cysteine ligase, the rate limiting enzyme in GSH synthesis [56]. In addition, preventive effects of β-carotene are associated with the β-carotene cleavage enzyme β-carotene 15,15′-monooxygenase (BCMO1) [57]. In the human macrophage cell line THP-1, β-carotene inhibited 7-ketocholesterol (7KC)-induced apoptosis by reducing expression levels of p53, p21, and Bax and inducing expression of AKT, Bcl-2, and Bcl-xL. Concomitantly, 7KC induced ROS generation with enhanced expression of NAD(P)H oxidase (NOX4). However, β-carotene blocked 7KC-induced ROS generation by inhibiting NOX4 [58]. Together these results indicate a possible antiarteriosclerotic action of β-carotene mediated

Carotenoids	Preventive effects	Mechanism of effects	Experiment procedure	Reference (author, issue year)
β-Carotene	Reverses the IL-1β-induced decrease in paraoxonase-1 expression	Induction of the CaMKKII pathway	*In vitro*	Yamagata et al., 2012
	Prevent the TNFα-induced decrease nitro-oxidative stress and interaction with monocytes	Prevention of induced, inflammation, decrease of ROS generation, increased NO/cGMP levels and reduces NF-κB–dependent adhesion molecule expression	*In vitro*	Di et al., 2012
Lycopene	Inhibited endothelin-1 expression and induces heme oxygenase-1	Block of ROS generation through NAD(P)H oxidase activity	*In vitro*	Sung et al., 2015
	Improved endothelium-dependent vasodilatation	Low C-reactive protein levels in CVD and health volunteer	*In vivo*	Gajendragadkar et al., 2014
	Increase endothelial function	Reduce oxidative stress, low C-reactive protein levels and decreased ICAM-1. VCAM-1	*In vivo*	Kim et al., 2011
	Reduce proinflammatory cytokine cascade	Inhibit TLR4 and NF-kappaB signaling pathway	*In vitro*	Wang et al., 2013
Astaxanthin	Protect against glucose fluctuation.	Reduced ROS generation	*In vitro*	Abdelzaher et al., 2016
	Ameliorative effect on endothelial dysfunction in streptozotocin-induced diabetes rats. Reduced serum oxLDL and aortic MDA. Reduced endothelium-dependent vasodilator with ACh.	Inhibition of the ox-LDL/LOX-1-eNOS pathway	*In vivo*	Zhao et al., 2011

ACh: acetylcholine; cGMP: cyclic GMP; CVD: cardiovascular disease; IL-1: interleukin-1; ICAM-1: intercellular adhesion molecule-1; LOX-1: lectin-like oxidized low density lipoprotein (LDL) receptor-1; MDA: malondialdehyde; NO: nitric oxide; oxLDL: oxidized low-density lipoprotein; TLR4: Toll-like receptor 4; TNFα: tumor necrosis factor-alpha; VCAM-1: vascular cell adhesion molecule-1.

Table 4. Preventive effect of carotenoids on vascular endothelial cells and macrophages.

through 7KC in human macrophages. β-Carotene also prevents expression of inflammatory genes such as inducible NO synthase (iNOS), cyclooxygenase-2 (COX2), TNF-α, and IL-1 in LPS-enhanced macrophages by inhibiting redox-related NF-κB activation [59].

6. Conclusions

This review examined the protective effects of carotenoids on CVD and the beneficial health effects of dietary carotenoids. Many studies indicated that carotenoids exhibit bioactivity in vascular endothelial cells. Carotenoids have antioxidant activity and appear to support and maintain normal vascular endothelial cell function. Future research may reveal new beneficial effects of carotenoids and help elucidate their preventive mechanisms in CVD.

Abbreviation

CVD	cardiovascular disease
ET-1	endothelin-1
eNOS	endothelial nitric oxide synthase
ERK	extracellular signal-regulated kinases
HDL	high-density lipoprotein
HO-1	heme oxygenase-1
ICAM-1	intercellular adhesion molecule-1
IL-6	interleukin-6
iNOS	inducible NO synthase
7KC	7-ketocholesterol
KLF2	Krüppel-like factor 2
LDL	low-density lipoprotein
LPS	lipopolysaccharide
MCP-1	monocyte chemoattractant protein-1
MMP	matrix metalloproteinase
NO	nitric oxide
ROS	reactive oxygen species
PON1	paraoxonase-1
PGC-1α	peroxisome proliferator activated receptor-γ coactivator
TLR	toll-like receptor
VCAM-1	vascular cell adhesion molecule-1

Author details

Kazuo Yamagata

Address all correspondence to: kyamagat@brs.nihon-u.ac.jp

Laboratory of Molecular Health Science of Food, Department of Food Science &; Technology, College of Bioresource Science, Nihon University (NUBS), Fujisawa, Kanagawa, Japan

References

[1] Pereira MA, O'Reilly E, Augustsson K, Fraser GE, Goldbourt U, Heitmann BL, Hallmans G, Knekt P, Liu S, Pietinen P, Spiegelman D, Stevens J, Virtamo J, Willett WC, Ascherio A: Dietary fiber and risk of coronary heart disease: A pooled analysis of cohort studies. Arch Intern Med. 2004;**164**:370–376.

[2] Dauchet L, Amouyel P, Hercberg S, Dallongeville J: Fruit and vegetable consumption and risk of coronary heart disease: A meta-analysis of cohort studies. J Nutr. 2006;**136**:2588–2593

[3] Van Duyn MA, Pivonka E: Overview of the health benefits of fruit and vegetable consumption for the dietetics professional: Selected literature. J Am Diet Assoc. 2000;**100**:1511–1521.

[4] Appel LJ, Moore TJ, Obarzanek E, Vollmer WM, Svetkey LP, Sacks FM, Bray GA, Vogt TM, Cutler JA, Windhauser MM, Lin PH, Karanja N: A clinical trial of the effects of dietary patterns on blood pressure. DASH collaborative research group. N Engl J Med. 1997;**336**:1117–1124.

[5] Bazzano LA, Serdula MK, Liu S: Dietary intake of fruits and vegetables and risk of cardiovascular disease. Curr Atheroscler Rep. 2003;**5**:492–499.

[6] Bruckdorfer KR: Antioxidants and CVD. Proc Nutr Soc. 2008;**67**:214–222. DOI: 10.1017/S0029665108007052.

[7] Sesso HD, Buring JE, Norkus EP, Gaziano JM: Plasma lycopene, other carotenoids, and retinol and the riskof cardiovascular disease in women. Am J Clin Nutr. 2004;**79**:47–53

[8] Ito Y, Kurata M, Suzuki K, Hamajima N, Hishida H, Aoki K: Cardiovascular disease mortality and serum carotenoid levels: A Japanese population-based follow-up study. J Epidemiol. 2006;**16**:154–160.

[9] Crews H, Alink G, Andersen R, Braesco V, Holst B, Maiani G, Ovesen L, Scotter M, Solfrizzo M, van den Berg R, Verhagen H, Williamson G: A critical assessment of some biomarker approaches linked with dietary intake. Br J Nutr. 2001;**86** Suppl 1:S5-S35.

[10] Stahl W, Sies H: Lycopene: A biologically important carotenoid for humans? Arch Biochem Biophys. 1996;**336**:1–9.

[11] Gomes-Rochette NF, Da Silveira Vasconcelos M, Nabavi SM, Mota EF, Nunes-Pinheiro DC, Daglia M, De Melo DF: Fruit as potent natural antioxidants and their biological effects. Curr Pharm Biotechnol. 2016;**17**:986–993.

[12] Bendich A, Olson JA: Biological actions of carotenoids. FASEB J. 1989;**3**:1927–1932.

[13] Krinsky NI, Johnson EJ: Carotenoid actions and their relation to health and disease. Mol Aspects Med. 2005;**26**:459–516.

[14] Wang Y, Gao Y, Yu W, Jiang Z, Qu J, Li K: Lycopene protects against LPS-induced proinflammatory cytokine cascade in HUVECs. Pharmazie. 2013;**68**:681–684.

[15] Hung CF, Huang TF, Chen BH, Shieh JM, Wu PH, Wu WB: In endothelial cells, lycopene inhibited TNF-alpha-induced NF-kappaB activation, ICAM-1 expression and monocyte-endothelial adhesion. Eur J Pharmacol. 2008;**586**:275–282. DOI: 10.1016/j.ejphar.2008.03.001.

[16] Zhu J, Wang CG, Xu YG: Lycopene attenuates endothelial dysfunction in streptozotocin-induced diabetic rats by reducing oxidative stress. Pharm Biol. 2011;**49**:1144–1149. DOI: 10.3109/13880209.2011.574707.

[17] Stahl W, Sies H: Bioactivity and protective effects of natural carotenoids. Biochim Biophys Acta. 2005;**1740**: 101–107.

[18] Voutilainen S, Nurmi T, Mursu J, Rissanen TH: Carotenoids and cardiovascular health. Am J Clin Nutr. 2006;**83**:1265–1271.

[19] Tang G: Vitamin A value of plant food provitamin A—Evaluated by the stable isotope technologies. Int J Vitam Nutr Res. 2014;**84** Suppl 1:25–29. DOI: 10.1024/0300-9831/a000183.

[20] van Het Hof KH, West CE, Weststrate JA, Hautvast JG: Dietary factors that affect the bioavailability of carotenoids. J Nutr. 2000;**130**:503–506.

[21] Gan Y, Tong X, Li L, Cao S, Yin X, Gao C, Herath C, Li W, Jin Z, Chen Y, Lu Z: Consumption of fruit and vegetable and risk of coronary heart disease: A meta-analysis of prospective cohort studies. Int J Cardiol. 2015;**183**:129–137. DOI: 10.1016/j.ijcard.2015.01.077.

[22] Jacques PF, Lyass A, Massaro JM, Vasan RS, D'Agostino RB Sr: Relationship of lycopene intake and consumption of tomato products to incident CVD. Br J Nutr. 2013;**110**: 545–551. DOI: 10.1017/S0007114512005417.

[23] Wang Y, Chung SJ, McCullough ML, Song WO, Fernandez ML, Koo SI, Chun OK: Dietary carotenoids are associated with cardiovascular disease risk biomarkers mediated by serum carotenoid concentrations. J Nutr. 2014;**144**:1067–1074. DOI: 10.3945/jn.113.184317.

[24] Hak AE, Ma J, Powell CB, Campos H, Gaziano JM, Willett WC, Stampfer MJ: Prospective study of plasma carotenoids and tocopherols in relation to risk of ischemic stroke. Stroke. 2004;**35**: 1584–1588.

[25] Osganian SK, Stampfer MJ, Rimm E, Spiegelman D, Hu FB, Manson JE, Willett WC: Vitamin C and risk of coronary heart disease in women. J Am Coll Cardiol. 2003;**42**:246–252.

[26] Hirvonen T, Virtamo J, Korhonen P, Albanes D, Pietinen P: Intake of flavonoids, carotenoids, vitamins C and E, and risk of stroke in male smokers. Stroke. 2000;**31**:2301–2306.

[27] Iwamoto T, Hosoda K, Hirano R, Kurata H, Matsumoto A, Miki W, Kamiyama M, Itakura H, Yamamoto S, Kondo K: Inhibition of low-density lipoprotein oxidation by astaxanthin. J Atheroscler Thromb. 2000;**7**: 216–222.

[28] Woodside JV, Young IS, McKinley MC: Fruit and vegetable intake and risk of cardiovascular disease. Proc Nutr Soc. 2013;**72**: 399–406. DOI: 10.1017/S0029665113003029.

[29] McEvoy CT, Wallace IR, Hamill LL, Hunter SJ, Neville CE, Patterson CC, Woodside JV, Young IS, McKinley MC: increasing fruit and vegetable intake has no dose–response effect on conventional cardiovascular risk factors in overweight adults at high risk of developing cardiovascular disease. J Nutr. 2015;**145**:1464–1471. DOI: 10.3945/jn.115.213090

[30] Wallace IR, McEvoy CT, Hunter SJ, Hamill LL, Ennis CN, Bell PM, Patterson CC, Woodside JV, Young IS, McKinley MC: Dose–response effect of fruit and vegetables on insulin resistance in people at high risk of cardiovascular disease: A randomized controlled trial. Diabetes Care. 2013;**36**: 3888–3896. DOI: 10.2337/dc13-0718.

[31] Hininger IA, Meyer-Wenger A, Moser U, Wright A, Southon S, Thurnham D, Chopra M, Van Den Berg H, Olmedilla B, Favier AE, Roussel AM: No significant effects of lutein, lycopene or beta-carotene supplementation on biological markers of oxidative stress and LDL oxidizability in healthy adult subjects. J Am Coll Nutr. 2001;**20**:232–238.

[32] Sesso HD, Buring JE, Norkus EP, Gaziano JM: Plasma lycopene, other carotenoids, and retinol and the risk of cardiovascular disease in men. Am J Clin Nutr. 2005;**81**: 990–997.

[33] Hak AE, Stampfer MJ, Campos H, Sesso HD, Gaziano JM, Willett W, Ma J: Plasma carotenoids and tocopherols and risk of myocardial infarction in a low-risk population of US male physicians. Circulation. 2003;**108**: 802–807.

[34] The Alpha-Tocopherol, Beta Carotene Cancer Prevention Study Group. The effect of vitamin E and beta carotene on the incidence of lung cancer and other cancers in male smokers. N Engl J Med. 1994;**330**:1029–1035.

[35] Hennekens CH, Buring JE, Manson JE, Stampfer M, Rosner B, Cook NR, Belanger C, LaMotte F, Gaziano JM, Ridker PM, Willett W, Peto R: Lack of effect of long-term supplementation with beta carotene on the incidence of malignant neoplasms and cardiovascular disease. N Engl J Med. 1996;**334**:1145–1149.

[36] Stangl V, Kuhn C, Hentschel S, Jochmann N, Jacob C, Bohm V, Frohlich K, Muller L, Gericke C, Lorenz M: Lack of effects of tomato products on endothelial function in human subjects: Results of a randomised, placebo-controlled cross-over study. Br J Nutr. 2011;**105**:263–267. DOI: 10.1017/S0007114510003284.

[37] Armoza A, Haim Y, Bashiri A, Wolak T, Paran E: Tomato extract and the carotenoids lycopene and lutein improve endothelial function and attenuate inflammatory NF-κB signaling in endothelial cells. J Hypertens. 2013;**31**: 521–529. DOI: 10.1097/HJH.0b013e32835c1d01.

[38] Xaplanteris P, Vlachopoulos C, Pietri P, Terentes-Printzios D, Kardara D, Alexopoulos N, Aznaouridis K, Miliou A, Stefanadis C: Tomato paste supplementation improves endothelial dynamics and reduces plasma total oxidative status in healthy subjects. Nutr Res. 2012;**32**:390–394. DOI: 10.1016/j.nutres.2012.03.011.

[39] Vilahur G, Cubedo J, Padro T, Casani L, Mendieta G, Gonzalez A, Badimon L: Intake of cooked tomato sauce preserves coronary endothelial function and improves apolipoprotein A-I and apolipoprotein J protein profile in high-density lipoproteins. Transl Res. 2015;**166**:44–56. DOI: 10.1016/j.trsl.2014.11.004.

[40] Sharma KD, Karki S, Thakur NS, Attri S: Chemical composition, functional properties and processing of carrot-a review. J Food Sci Technol. 2012;**49**:22–32. DOI: 10.1007/s13197-011-0310-7.

[41] Potter AS, Foroudi S, Stamatikos A, Patil BS, Deyhim F: Drinking carrot juice increases total antioxidant status and decreases lipid peroxidation in adults. Nutr J. 2011 Sep 24;**10**:96. DOI: 10.1186/1475-2891-10-96.

[42] Poudyal H, Panchal S, Brown L: Comparison of purple carrot juice and β-carotene in a high-carbohydrate, high-fat diet-fed rat model of the metabolic syndrome. Br J Nutr. 2010;**104**:1322–1332. DOI: 10.1017/S0007114510002308.

[43] Di Tomo P, Canali R, Ciavardelli D, Di Silvestre S, De Marco A, Giardinelli A, Pipino C, Di Pietro N, Virgili F, Pandolfi A: β-Carotene and lycopene affect endothelial response to TNF-α reducing nitro-oxidative stress and interaction with monocytes. Mol Nutr Food Res. 2012;**56**:217–227. DOI:10.1002/mnfr.201100500.

[44] Sung LC, Chao HH, Chen CH, Tsai JC, Liu JC, Hong HJ, Cheng TH, Chen JJ: Lycopene inhibits cyclic strain-induced endothelin-1 expression through the suppression of reactive oxygen species generation and induction of heme oxygenase-1 in human umbilical vein endothelial cells. Clin Exp Pharmacol Physiol. 2015;**42**: 632–639. DOI: 10.1111/1440-1681.12412.

[45] Lee SJ, Bai SK, Lee KS, Namkoong S, Na HJ, Ha KS, Han JA, Yim SV, Chang K, Kwon YG, Lee SK, Kim YM: Astaxanthin inhibits nitric oxide production and inflammatory gene expression by suppressing I (kappa)B kinase-dependent NF-kappaB activation. Mol Cells. 2003;**16**:97–105.

[46] Sies H, Stahl W, Sundquist AR: Antioxidant functions of vitamins.Vitamins E and C, beta-carotene, and other carotenoids. Ann N Y Acad Sci. 1992;**30;**669:7–20.

[47] Shimokawa H: Primary endothelial dysfunction: atherosclerosis. J Mol Cell Cardiol. 1999;**31**:23–37.

[48] Kim JY, Paik JK, Kim OY, Park HW, Lee JH, Jang Y, Lee JH: Effects of lycopene supplementation on oxidative stress and markers of endothelial function in healthy men. Atherosclerosis. 2011; **215**:189–195. DOI: 10.1016/j.atherosclerosis.2010.11.036.

[49] Gajendragadkar PR, Hubsch A, Maki-Petaja KM, Serg M, Wilkinson IB, Cheriyan J: Effects of oral lycopene supplementation on vascular function in patients with cardiovascular disease and healthy volunteers: A randomised controlled trial. PLoS One. 2014;**9**:e99070. DOI: 10.1371/journal.pone.0099070. eCollection 2014.

[50] Mackness MI, Arrol S, Abbott C, Durrington PN: Protection of low-density lipoprotein against oxidative modification by high-density lipoprotein associated paraoxonase. Atherosclerosis. 1993;**104**:129–135.

[51] Yamagata K, Tanaka N, Matsufuji H, Chino M: β-Carotene reverses the IL-1β-mediated reduction in paraoxonase-1 expression via induction of the CaMKKII pathway in human endothelial cells. Microvasc Res. 2012;**84**: 297–305. DOI: 10.1016/j.mvr.2012.06.007.

[52] Abdelzaher LA, Imaizumi T, Suzuki T, Tomita K, Takashina M, Hattori Y: Astaxanthin alleviates oxidative stress insults-related derangements in human vascular endothelial cells exposed to glucose fluctuations. Life Sci. 2016;**150**:24–31. DOI: 10.1016/j.lfs.2016.02.087.

[53] Zhao ZW, Cai W, Lin YL, Lin QF, Jiang Q, Lin Z, Chen LL: Ameliorative effect of astaxanthin on endothelial dysfunction in streptozotocin-induced diabetes in male rats. Arzneimittelforschung. 2011;**61**:239–246. DOI: 10.1055/s-0031-1296194.

[54] Regnier P, Bastias J, Rodriguez-Ruiz V, Caballero-Casero N, Caballo C, Sicilia D, Fuentes A, Maire M, Crepin M, Letourneur D, Gueguen V, Rubio S, Pavon-Djavid G: Astaxanthin from haematococcus pluvialis prevents oxidative stress on human endothelial cells without toxicity. Mar Drugs. 2015;**13**: 2857–2874. DOI: 10.3390/md13052857.

[55] Lo HM, Chen CL, Yang CM, Wu PH, Tsou CJ, Chiang KW, Wu WB: The carotenoid lutein enhances matrix metalloproteinase-9 production and phagocytosis through intracellular ROS generation and ERK1/2, p38 MAPK, and RARβ activation in murine macrophages. J Leukoc Biol. 2013;93:723–735. DOI: 10.1189/jlb.0512238

[56] Akaboshi T, Yamanishi R: Certain carotenoids enhance the intracellular glutathione level in a murine cultured macrophage cell line by inducing glutamate-cysteine-ligase. Mol Nutr Food Res. 2014;58: 1291–1300. DOI: 10.1002/mnfr.201300753.

[57] Zolberg Relevy N, Bechor S, Harari A, Ben-Amotz A, Kamari Y, Harats D, Shaish A: The inhibition of macrophage foam cell formation by 9-cis β-carotene is driven by BCMO1 activity. PLoS One. 2015;**10**: e0115272. DOI: 10.1371/journal.pone.0115272

[58] Palozza P, Simone R, Catalano A, Boninsegna A, Bohm V, Frohlich K, Mele MC, Monego G, Ranelletti FO: Lycopene prevents 7-ketocholesterol-induced oxidative stress, cell

cycle arrest and apoptosis in human macrophages. J Nutr Biochem. 2010;**21**:34–46. DOI:10.1016/j.jnutbio.2008.10.002.

[59] Bai SK, Lee SJ, Na HJ, Ha KS, Han JA, Lee H, Kwon YG, Chung CK, Kim YM: Beta-carotene inhibits inflammatory gene expression in lipopolysaccharide-stimulated macrophages by suppressing redox-based NF-kappaB activation. Exp Mol Med. 2005;**37**: 323–334.

Section 2

Biotechnological Applications

Chapter 8

Biotechnological Production of Carotenoids and Their Applications in Food and Pharmaceutical Products

Ligia A. C. Cardoso, Susan G. Karp,
Francielo Vendruscolo, Karen Y. F. Kanno,
Liliana I. C. Zoz and Júlio C. Carvalho

Additional information is available at the end of the chapter

http://dx.doi.org/10.5772/67725

Abstract

Pigments can be divided into four categories: natural, nature-identical, synthetic, and inorganic colors. Artificial colorants are the most used in food and pharmaceutical industries because of their advantages related to color range, price, resistance to oxygen degradation, and solubility. However, many natural pigments present health-promoting activities that make them an interesting option for human use and consumption. Natural colorants are derived from sources such as plants, insects, and microorganisms. Carotenoids are natural pigments with important biological activities, such as antioxidant and pro-vitamin A activity, that can be either extracted from plants and algae or synthesized by various microorganisms, including bacteria, yeasts, filamentous fungi, and microalgae. Advantages of microbial production include the ability of microorganisms to use a wide variety of low cost substrates, the better control of cultivation, and the minimized production time. After fermentation, carotenoids are usually recovered by cell disruption, solvent extraction, and concentration. Subsequent purification steps are followed depending on the application. The most prominent industrial applications of carotenoids, considering their health benefits, are in the food, feed, and pharmaceutical industries.

Keywords: biotechnology, natural pigments, microbial carotenoids, downstream, industrial applications

1. Introduction

Color has a great influence on the appearance, processing, and acceptance of food products, textiles and pharmaceutical products. The first quality impact by which consumers make the decision to purchase a product is its visual appearance.

Food colorants can be divided into four categories: natural, nature-identical, synthetic, and inorganic colors [1]. The production of synthetic coloring agents and other chemicals used as food additives is under increasing pressure due to a renewed interest in the use of natural products and the strong interest in minimizing the use of chemical processes [2]. Since the number of permitted synthetic colorants has decreased because of undesirable toxic effects including mutagenicity and potential carcinogenicity, interest focuses on the development of food grade pigments from natural sources [3–5].

Natural pigments are derived from sources such as plants, insects, and microorganisms. Algae and microalgae, bacteria, fungi, and yeasts are organisms commonly found in nature that can produce natural pigments in different color spectra, such as violacein, phycocyanin, monascins, flavins, quinones, and carotenoids.

Carotenoids represent one of the most important groups of natural pigments, they are responsible for the yellow, orange, red, and purple colors in a wide variety of plants, animals, and microorganisms [6]. They are lipid-soluble, commercially, and biotechnologically significant pigments produced from various organisms such as plants [7], algae and microalgae [8–12], bacteria [13–15], fungi [16–20], and yeasts [4, 21–25].

Pigments from natural sources have been obtained since long time ago, and their attractiveness has increased due to the toxicity problems caused by the synthetic pigments [26–28]. Carotenoids are obtained industrially by chemical synthesis or extraction from plants or algae; however, there has been an increasing interest in biotechnological processes for carotenoids production [29]. The pigments from microbial sources are a good alternative to obtain natural colorants for industrial uses.

The biotechnological production of carotenoids has advantages related to the diversity of microorganisms in nature, versatility in the use of substrates and agro-industrial wastes and the possibility to control operating conditions such as pH, temperature, dissolved oxygen, and light intensity; also, biomass from other bioprocesses can be submitted to the extraction of carotenoids. The production of microbial carotenoids has become a potential alternative for the replacement of artificial pigments, even with technological, economic, and legislation limitations.

Studies have demonstrated that carotenoids play an essential role for the maintenance of living bodies. In plants, carotenoids play an important role in photosynthesis, acting as light-harvesting pigments and protectors against photo-oxidation. In foods, carotenoids confer yellow, orange, or red color, serve as precursors of aroma compounds, and, as natural antioxidants, may help to extend the shelf-life [30, 31]. In humans, carotenoids have been associated with the reduction of the risk of developing chronic diseases such as cancer, cardiovascular diseases, high levels of cholesterol, cataract, and macular degeneration, aside from the pro-vitamin A activity of some of these compounds [31–33]. This is important because in the developed world,

http://dx.doi.org/10.5772/67725

as life expectancy increases and the birth rate declines, the demand for solutions focusing on longevity and life quality increases too. The number of people aged >60 years is expected to account for approximately one-fifth of the world's population by 2050 [34].

2. Biotechnological production of carotenoids

2.1. Carotenoids diversity

Carotenoids are lipid-soluble pigments, colored from yellow to red, with a basic structure consisting in a tetraterpene with a series of conjugated double bonds. They can have only carbon and hydrogen in their structures or have one or more oxygen atoms, being classified as xanthophylls. The majority of carotenoids are C_{40} terpenoids, which act as membrane-protective antioxidants scavenging O_2 and peroxyl radicals [35].

There are more than 700 types of carotenoids described and only about 50 are precursors of vitamin A. Carotenoids can reduce risks for degenerative diseases such as cancer, cardiovascular diseases, macular degeneration, and cataract. The biological activities, specially the antioxidant properties, depend on their chemical structure: number of conjugated double bonds, structural end-groups, and oxygen-containing substituents [36].

Carotenoids occur in photosynthetic systems of higher plants, algae, and phototrophic bacteria. In plants, carotenoids are embedded in the membranes of chloroplasts and chromoplasts. The colors of these pigments are masked by chlorophyll, but they contribute to the bright colors of many flowers and fruits [37].

Nonphotosynthetic organisms, as some bacteria and fungi, present carotenoids as protectors against photo-oxidative damage, a way of protection in growth conditions with light and abundant air. The main carotenoids produced by fungi are β-carotene, torulene, torularhodin, and astaxanthin [38]. Bacteria have been reported as producers of cantaxanthin mainly. The microalgae are producers of lutein, β-carotene, and astaxanthin [35].

Animals usually present carotenoids provenient from their diet. Marine animals that feed on algae or on products rich in carotenoids may exhibit the coloration of these pigments, as the salmon fish. The color of the feathers of some birds also comes from a diet rich in carotenoids, as flamingos [39].

The industrial production of carotenoids by plants is dependent on the season and geographic variability, and these cannot always be controlled. The chemical synthesis of carotenoids generates wastes that can cause damage to the environment and resistance by the consumers. Because of this, the biotechnological resources are becoming more interesting. The microbial production of carotenoids can be performed using low-cost substrates or substrates that are residues from industrial processes, like molasses, resulting in lower costs of production [40]. All conditions of this kind of production can be controlled and optimized, especially knowing the metabolic route of each microorganism utilized.

Carotenoids are intracellular products, and a process to increase their accessibility at the downstream stage is necessary. The techniques most used combine physical and chemical methods like maceration and contact with organic solvents [4].

2.2. Main carotenoid biosynthesis pathways

Carotenoids are usually produced from the building blocks geranyl geranyl diphosphate (GGPP) and farnesyl diphosphate (FPP), like other secondary metabolites such as sesquiterpenoids and steroids. The most common pathway is the condensation of 2 GGPP units into prephytoene diphosphate and then to phytoene, a 40-carbon polyunsaturated precursor which is colorless. This precursor is converted into lycopene and then into several derived carotenoids such as β-carotene and oxidized derivatives such as lutein. The condensation of two units of FPP leads to 30-carbon precursors that are converted to steroids or apocarotenoids such as staphyloxanthin [41, 42]. Apocarotenoids can also be produced by oxidative cleavage of carotenoids. **Figure 1** presents a simplified carotenoid biosynthesis pathway.

Most carotenoids present maximal absorption in the violet to green region of the visible spectrum, so these substances appear as red to yellow pigments. **Table 1** shows the carotenoids with permitted food use according to the Food and Drug Administration (FDA) and the Food and Agriculture Organization (FAO).

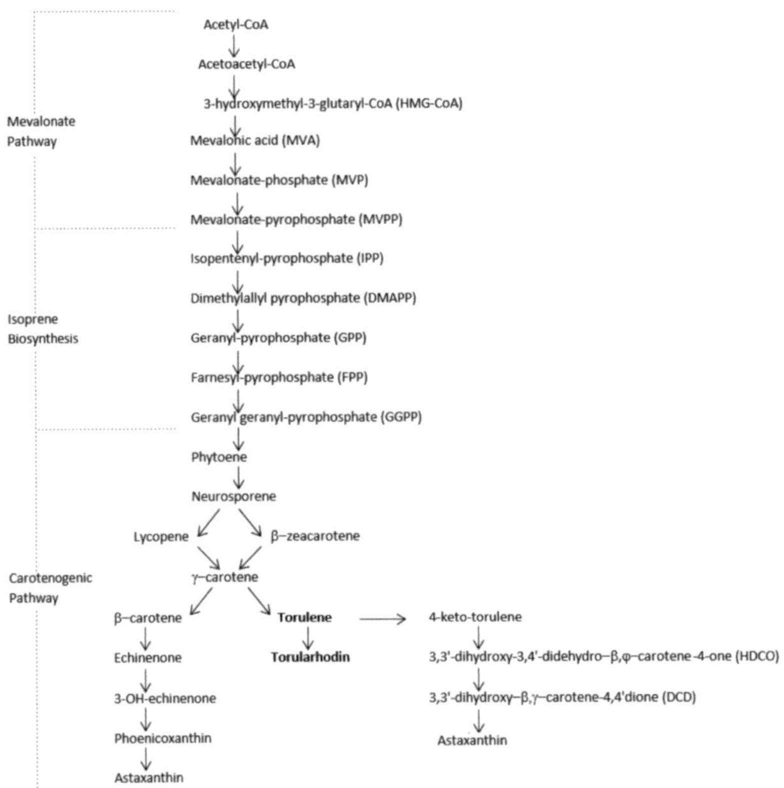

Figure 1. Biosynthesis pathways of common carotenoids. Source: Adapted from Ref. [38] with permission.

http://dx.doi.org/10.5772/67725

Additive/source	Color	Main component	International Numbering System
Algae meal, dried	Green to red	Mixture of carotenoids, xanthophyll, and chlorophylls	
Astaxanthin and astaxanthin dimethyldisuccinate (several microorganisms)	Orange-red	Astaxanthin	
β-apo-8′-carotenal (from carrot oil)	Reddish orange	All-trans-β-apo-8′-carotenal	160e, 160f
β-carotene, synthetic and natural: from vegetables, *Blakeslea trispora* and *Dunaliella salina*	Orange	β-carotene	160a(i), 160a(ii)
Canthaxanthin; most of the pigment used in feeds is synthetic	Orange pink	β-carotene-4,4′-dione (canthaxanthin)	161g
Carrot oil	Orange to yellow		160a(ii)
Gardenia red and yellow	Red, yellow	Crocin, crocetin	
Haematococcus algae meal	Orange-red	Astaxanthin	
Lycopene, tomato extract (i) or concentrate (ii) or from *Blakeslea trispora* (iii)	Red	Lycopene	160d
Lutein (bras), from marigold oleoresin		Lutein	161b
Marigold color	Yellow	Lutein	
Orchil dyes	Red	Orcein	
Paprika and paprika oleoresin	Red	Capsanthin, capsorubin	160c
Paracoccus pigment	Red	Astaxanthin	
Phaffia yeast	Orange-red	Astaxanthin esters	
Saffron	Yellow to orange	α-crocin	164
Tagetes (Aztec marigold) meal and extract	Yellow to orange	Lutein	161b

Sources: Compiled from the FDA Color Additive Status List [http://www.fda.gov/ForIndustry/ColorAdditives/ColorAdditiveInventories/ucm106626.htm] and from the Combined Compendium of Food Additive Specifications [ftp://ftp.fao.org/docrep/fao/009/a0691e/a0691e00a.pdf].

Table 1. Carotenoids and carotenoid-rich products used as food color additives.

2.3. Carotenoid sources

The most common sources for natural carotenoids for food and cosmetic use are plants, although microorganism biomass is becoming more common as a source for these substances. **Table 2** illustrates some commercial sources for microorganism-based carotenoids.

Microorganism	Molecule	Culture medium*	X_{max} (g/L)	P_{max} (mg/L)	Conc. (mg/g)**	μ_x (h^{-1})	References
Blakeslea trispora (fungus)	β-carotene	Corn steep liquor	20	800	40	0.022	[43]
Blakeslea trispora	β-carotene	Whey	8	1360	170	0.023	[44]
Sporobolomyces roseus (yeast)	β-carotene	Reconstituted whey	4.71	2.58	0.55	–	[40]
Rhodotorula glutinis (yeast)	β-carotene	Potato extract	5.70	1.08	0.19	–	[40]
Dietzia natronolimnaea (bacterium)	Canthaxanthin	Whey	3.29	2.87	0.87	0.020	[45]
Phaffia rhodozyma (yeast)	Astaxanthin	Cassava residues	8.6	2.98	0.35	0.060	[46]
Sporobolomyces ruberrimus (yeast)	Torularhodine	Technical glycerol	30	3.7	0.12	0.040	[47]
Chlorella zofingiensis (microalga)	Astaxanthin	BBM with glucose	10.2	–	1	0.031	[48]
Coelastrella striolata (microalga)	Canthaxanthin Astaxanthin β-carotene	BBM	2.7	–	47.5 1.5 7	0.30	[49]
Coccomyxa onubensis (microalga)	β-carotene Lutein	K9	1.6	–	2.88 6.48	0.50	[50]
Haematococcus pluvialis (microalga)	Astaxanthin	BBM	2.2	–	13.5	–	[51]
Chlorella zofingiensis	Astaxanthin	Bristol, modified	10	–	1.25	0.043	[52]
Dunaliella salina (microalga)	β-carotene	f2	–	–	14***	0.55	[53]
Haematococcus pluvialis	Astaxanthin	Standard	3	–	12–15	0.56	[54]
Muriellopsis sp. (microalga)	Lutein	Arnon, modified	5.37	–	6.51	0.17–0.23	[55]
Haematococcus pluvialis (wild-type) *Haematococcus pluvialis* (mutant)	Astaxanthin	NIES medium	1.6 2.25	– –	47.62 54.78	0.07 0.08	[56]
Paracoccus carotinifaciens (bacterium)	Astaxanthin Canthaxanthin	Glucose and peptone based	–	–	25–40	–	[57]

*Except where specified, these are mineral-based media. Recipes may be found at UTEX, SAG, or CCMP collections web sites.

**Milligrams of carotenoids per gram of biomass.

***Estimated. The original reference reports 28.1 mg/L carotenoids.

X_{max}—maximum biomass concentration; P_{max}—maximum carotenoids concentration; μ_X—biomass production rate.

Source: Adapted from Ref. [58].

Table 2. Main sources for concentrated carotenoids.

2.4. General downstream operations for carotenoid production

Carotenoids are nonpolar molecules that accumulate intracellularly in plant tissues and microorganisms. Therefore, the production usually consists in a biomass pretreatment that may accelerate the dissolution of these substances, followed by a solid-liquid extraction (leaching) with a suitable, low-polarity solvent. The resulting solution can be a final product, can be desolventized, and can be further purified, depending on the use intended for the extract. **Figure 2** illustrates the main steps in the production of carotenoids.

The first step in carotenoid production is the pretreatment of the raw biomass, usually by drying and milling. Drying is convenient because it reduces the weight of the material to be processed, facilitates the access for solvents to the biomass, and reduces contaminants that could be extracted in water micelles with the solvent. The milling step is also important because it increases the surface area of the biomass matrix, facilitating contact with the solvent. In the case of tough-walled organisms, chemical or mechanical cell disruption may be done prior to drying. Fine milling of the dry biomass is less common.

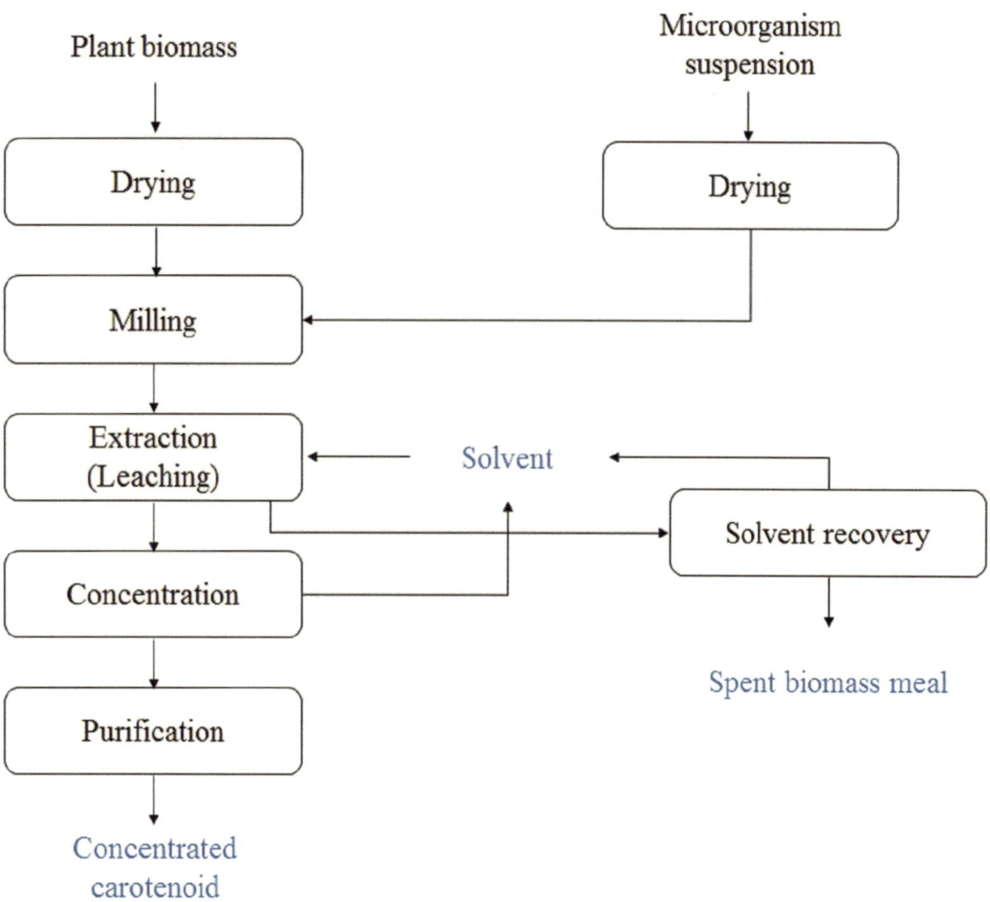

Figure 2. Main steps in carotenoids production.

The dry biomass is then extracted using a nonpolar solvent such as hexane or a vegetable oil, for the dissolution of carotenoids. A higher polarity solvent such as acetone can be used for the extraction of xanthophylls. In both cases, lipids are extracted in the mix. This extraction is an equilibrium operation; therefore, the final concentration in the solvent affects the extraction efficiency. Following extraction, the solution containing carotenoids must be concentrated and desolventized. This is why low boiling point solvents, which are easy to evaporate, are more common extractants than oils.

The carotenoids in the concentrated extract may be purified or not, depending on the intended use. For example, β-carotene that will be used as a vitamin A precursor must be purified, while paprika oleoresin is a mixture of carotenoids used mainly as a color and flavor additive and needs no further purification. In general, for carotenoids used as color additives, it is enough to concentrate the extract because (1) the tinctorial strength of the molecules is large—therefore, the additive is added at a low concentration to the formulated product and (2) the sources used are generally regarded as safe (GRAS), and the molecules extracted with the carotenoid are harmless in the concentrations used.

In the case of purified carotenoids, the operations to be used—adsorption, chromatography, crystallization, etc.—depend largely on the properties of the target molecule and the contaminants in the mixture, such as melting point, polarity, solubility, etc. All sorts of nonpolar compounds are extracted with the solvent, such as neutral and slightly polar lipids, steroids, and waxes. The differences in the properties of the carotenoid and the contaminants will be explored in the purification strategy.

Following extraction and purification, the carotenoid must be formulated for further application. This formulation will also depend on the intended use. The formulation may be as simple as adding an antioxidant such as butylated hydroxytoluene (BHT) or butylated hydroxyanisole (BHA) to the extract or may be more complex, such as emulsifying the carotenoid as an oil-in-water product for use in polar matrixes such as juices.

3. Industrial application of carotenoids as additives in food, feed and pharmaceutical products

Because of the rising of health concerns by consumers, the demand for carotenoids as natural coloring products is growing. Beta-carotene, astaxanthin, canthaxanthin, lycopen, and lutein are the most required and valuable carotenoids, and they are currently used by the food, feed, and cosmetic industries (**Table 3**). The use of carotenoids is regulated by the legislation of each country that specifies the source, purity, product, and quantities of the colorant that can be used [59].

According to BBC Research [65], the carotenoid global market in 2014 was of US$ 1.5 billion, this value is increasing year by year and is expected to reach US$ 1.8 billion in 2019, with an annual growth rate of 3.9%. Beta-carotene, the carotenoid of highest value, had a global market of US$ 233 million in 2010, which is expected to reach US$ 309 million by 2018. Astaxanthin, due to its powerful antioxidant activity, is the third carotenoid in terms of high added value, with a global market size of US$ 225 million, estimated to increase to US$ 253 million by 2018.

http://dx.doi.org/10.5772/67725

Carotenoid	Color	Application	Activities	References
Lutein	Yellow	Poultry feed; functional nutrient	Antioxidant	[60]
Canthaxanthin	Orange	Poultry feed; fish feed; cosmetic	Antioxidant, anticancer	[61]
Lycopene	Red	Supplement in functional foods; additive in cosmetics	Antioxidant, anticancer	[62]
β-carotene	Orange-red	Nutraceutical; cosmetic; animal feed industries	Antioxidant, anticancer, precursor of vitamin A	[63]
Astaxanthin	Pink-red	Fish feed; cosmetic industry	Antioxidant, photoprotectant, anticancer, anti-inflammatory	[64]

Table 3. Carotenoids' colors, applications and biological activities.

3.1. Importance and use of carotenoids in food products

Commercial food products using carotenoids are expanding, and the greatest demand is in the Asian continent. The pigment is extracted from microalgae such as *Chlorella, Dunaliella, Haematococcus* [66, 67], from the cyanobacterium *Spirulina* [68], and from the fungus *Monascus* [69].

In Asia, the production red *koji* dates of hundreds of years and uses the fermentation of rice by *Monascus* to produce the typical reddish color. These red pigments are also used as food colorants for wine, red soy cheese, meat, and by-products of meat and fish [26]. The French cheese named *vieux-pan* contains the carotenoid produced by *Brevibacterium linens* due to its orange-red-brown color that improves the sensory quality of the product [70]. In Russia, infant formulas are enriched with natural pigments such as lutein, which is present in breast milk, in order to improve children's health [71].

Nutraceutical food products have also been applied in bakery products and pasta. In Japan, *Undaria pinnatifida* (wakame), an edible seaweed rich in fucoxanthin, is commercialized as an ingredient for pasta [72]. In India, a pasta containing fucoxanthin as an ingredient to improve its biofunctional and nutritional qualities was developed [73].

3.2. Importance and applications of carotenoids in the pharmaceutical industry

Besides the use of nutraceutical foods as a form of prevention and treatment of diseases, the administration of the bioactive compounds in their concentrated form is also a possibility for promoting health. The transport of carotenoids occurs from the intestinal mucosa to the blood vessels carried by lipoproteins [74]. Carotenoids functional properties are related to reactions such as oxidation, reduction, hydrogen abstraction, and addition in biological membranes, and their antioxidant power is fundamental for cell protection against free radicals and singlet oxygen formed in tissues [75].

Some carotenoids are precursors of vitamins, and they also present activities such as anti-inflammatory, antioxidant, immunomodulatory, anticancer, for cardiovascular therapy and neurodegenerative diseases [76], and anti-obesity [77]. The carotenoids included as pro-vitamin A are β-carotene, α-carotene, and cryptoxanthin. Vitamin A is an essential nutrient for operation and maintenance of biological functions including vision, reproduction, and immunity [78]. Beta-carotene is present in blood and tissues, which is associated with antioxidant activity and concomitantly with other carotenoids or antioxidants can enhance their activity against free radicals. However, it can bring health risk at high doses [79].

Carotenoids, acting as antioxidants eliminating free radicals, can modulate the risk of developing chronic diseases by inhibiting reactions mediated by reactive oxygen species (ROS). Reactive species are produced during cellular metabolism as a defense to infectious and chemical agents that may cause damage to DNA, proteins, and tissues, contributing to the development of chronic diseases such as diabetes, Parkinson's, Alzheimer's, cardiovascular diseases, and cancer [80].

In addition to the antioxidant properties, carotenoids exhibit anti-inflammatory activities owing to the protective effects of phytochemicals such as lutein and astaxanthin. Astaxanthin has been shown to inhibit the production of pro-inflammatory mediators such as nitric oxide (NO) in macrophages, to increase the level of inflammatory cytokines, and to reduce oxidative stress. Neuroprotective effect, reduced neuroinflammation, improvement of insulin signals, and reduction of lipid levels were also verified [81].

Inhibition of cell proliferation of colon cancer cells by the use of *Neochloris oleoabundans* carotenoids was observed, enabling its use as a functional food additive or nutraceutical with potential for the prevention of colon cancer [82]. Beta-carotene, astaxanthin, and capsanthin demonstrated antiproliferative effects on leukemic K562 cells [83]. Studies indicated that the simultaneous use of different carotenoids was efficient against liver cancer. Patients were administered with β-cryptoxanthin-enriched mandarin orange juice and capsules of a carotenoids mixture-containing lutein, β-cryptoxanthin, lycopene, zeaxanthin, and fucoxanthin. Analyses of DNA array and protein-antibody array showed that the carotenoids interferred in the induction of genes such as p16 and p73 [84].

4. Conclusion and final remarks

There are many advantages related to the use of carotenoids instead of artificial pigments in food products and for pharmaceutical applications. Their biological properties such as antioxidant, anti-inflammatory, antitumoral, and pro-vitamin A activities contribute to the quality of the product and to the consumer's health. Among the production strategies, microbial synthesis is considered advantageous, and the downstream techniques usually involve cell disruption, solvent extraction, concentration, and purification, when necessary. Several researches have proved the beneficial effects of carotenoids on health, so they can meet the demand for solutions focusing on longevity and life quality.

Author details

Ligia A. C. Cardoso[1]*, Susan G. Karp[2], Francielo Vendruscolo[3], Karen Y. F. Kanno[1], Liliana I. C. Zoz[2] and Júlio C. Carvalho[2]

*Address all correspondence to: ligiacardoso@up.edu.br

1 Positivo University, Curitiba, Brazil

2 Federal University of Paraná, Curitiba, Brazil

3 Federal University of Goiás, Goiás, Brazil

References

[1] Aberoumand A. A review article on edible pigments properties and sources as natural biocolorants in foodstuff and food industry. World Journal of Dairy & Food Sciences. 2011;**6**:71-78.

[2] Domínguez-Espinosa RM, Webb C. Submerged fermentation in wheat substrates for production of *Monascus* pigments. World Journal of Microbiology and Biotechnology. 2003;**19**:329-336. DOI: 10.1023/A:1023609427750

[3] Sabater-Vilar M, Maas RFM, Fink-Gremmels J. Mutagenicity of commercial *Monascus* fermentation products and the role of citrinin contamination. Mutation Research. 1999;**444**:7-16. DOI: 10.1016/S1383-5718(99)00095-9

[4] Pennacchi MGC, Rodrígues-Fernández DE, Vendruscolo F, Maranho LT, Marc I, Cardoso LAC. A comparison of cell disruption procedures for the recovery of intracellular carotenoids from *Sporobolomyces ruberrimus* H110. International Journal of Applied Biology and Pharmaceutical Technology. 2015;**6**:136-143.

[5] Vendruscolo F, Bühler RMM, Carvalho JC, Oliveira D, Moritz DE, Schmidell W, Ninow JL. *Monascus*: A reality on the production and application of microbial pigments. Applied Biochemistry and Biotechnology. 2016;**178**:211-223. DOI: 10.10007/s12010-015-1880-z

[6] Oliver J, Palou A. Chromatographic determination of carotenoids in foods. Journal of Chromatography A. 2000;**881**:543-555.

[7] Hanson P, Yang RY, Chang LC, Ledesma L, Ledesma D. Mint: Carotenoids, ascorbic acid, minerals, and total glucosinolates in choysum (*Brassica rapa* cvg. *parachinensis*) and kailaan (*B. oleraceae* Alboglabra group) as affected by variety and wet and dry season production. Journal of Food Composition and Analysis. 2011;**24**:950-962. DOI: 10.1016/j.jfca.2011.02.001

[8] Rodrigues DB, Flores EMM, Barin JS, Mercadante AZ, Jacob-Lopes E, Zepka LQ. Production of carotenoids from microalgae cultivated using agroindustrial wastes. Food Research International. 2014;**65**:144-148. DOI: 10.1016/j.foodres.2014.06.037

[9] Přibyl P, Cepák V, Kaštánek P, Zachleder V. Elevated production of carotenoids by a new isolate of *Scenedesmus* sp. Algal Research. 2015;**11**:22-27. DOI: 10.1016/j.algal.2015.05.020

[10] Chen L, Zhang L, Liu T. Concurrent production of carotenoids and lipid by a filamentous microalga *Trentepohlia arborum*. Bioresource Technology. 2016;**214**:567-573. DOI: 10.1016/j.biortech.2016.05.017

[11] Liu J, Mao X, Zhou W, Guarnieri MT. Simultaneous production of triacylglycerol and high-value carotenoids by the astaxanthin-producing oleaginous green microalga *Chlorella zofingiensis*. Bioresource Technology. 2016;**214**:319-327. DOI: 10.1016/j.biortech.2016.04.112

[12] Tsai HP, Chuang LT, Chen CNN. Production of long chain omega-3 fatty acids and carotenoids in tropical areas by a new heat-tolerant microalga *Tetraselmis* sp. DS3. Food Chemistry. 2016;**192**:682-690. DOI: 10.1016/j.foodchem.2015.07.071

[13] Fang CJ, Ku KL, Lee MH, Su NW. Mint: Influence of nutritive factors on C_{50} carotenoids production by *Haloferax mediterranei* ATCC 33500 with two-stage cultivation. Bioresource Technology. 2010;**101**:6487-6493. DOI: 10.1016/j.biortech.2010.03.044

[14] Peter-Wendisch P, Götker S, Heider SAE, Reddy K, Nguyen AQ, Stansen KC, Wendisch VF. Engineering biotin prototrophic *Corynebacterium glutamicum* strains for amino acid, diamine and carotenoid production. Journal of Biotechnology. 2014;**192**:346-354. DOI: 10.1016/j.jbiotec.2014.01.023

[15] Autenrieth C, Ghosh R. Random mutagenesis and overexpression of rhodopin-3,4-desaturase allows the production of highly conjugated carotenoids in *Rodospirillum rubrum*. Archives of Biochemistry and Biophysics. 2015;**572**:134-141. DOI: 10.1016/j.abb.2015.01.023

[16] Goodwin TW. Fungal carotenoids. Botanical Review. 1952;**18**:291-316.

[17] El-Jack M, Mackenzie A, Bramley PM. The photoregulation of carotenoid biosynthesis in *Aspegillus giganteus* mut. *alba*. Planta. 1998;**174**:59-66.

[18] Denter J, Rehm HJ, Bisping B. Changes in the contents of fat-soluble vitamins and provitamins during tempo fermentation. International Journal of Food Microbiology. 1998;**45**:129-134.

[19] Iturriaga EA, Papp T, Breum J, Arnau J, Eslava AP. Strain and culture conditions improvement for b-carotene production in *Mucor*. In: Microbial Processes and Products, Methods in Biotechnology series. 1st ed. Humana Press; 2005. pp. 239-256. DOI: 10.1385/1-59259-847-1:239.

[20] Csernetics A, Nagy G, Iturriaga EA, Szekeres A, Eslava AP, Vágvölgyi C, Papp T. Expression of three isoprenoid biosynthesis genes and their effects on the carotenoid production of the zygomycete *Mucor circinelloides*. Fungal Genetics and Biology. 2011;**48**:696-703. DOI: 10.1016/j.fgb.2011.03.006

[21] Valduga E, Ribeiro AHR, Cence K, Coilet R, Tiggemann L, Zeni J, Toniazzo G. Carotenoids production from a newly isolated *Sporidiobolus pararoseus* strain using agro-industrial substrates. Biocatalysis and Agricultural Biotechnological. 2014;**3**:207-213. DOI: 10.1016/j.bcab.2013.10.001

[22] Dias C, Sousa S, Caldeira J, Reis A, Silva TL. New dual-stage pH control fed-batch cultivation strategy for the improvement of lipids and carotenoids production by the red yeast *Rhodosporidium toruloides* NCYC 921. Bioresource Technology. 2015;**189**:309-318. DOI: 10.1016/j.biortech.2015.04.009

[23] Cardoso LAC, Jäckel S, Karp SG, Framboisier X, Chevalot I, Marc I. Improvement of *Sporobolomyces ruberrimus* carotenoids production by the use of raw glycerol. Bioresource Technology. 2016;**200**:374-379. DOI: 10.1016/j.biortech.2015.09.108

[24] Odoñez MC, Raftery JP, Jaladi T, Chen X, Kao K, Karim MN. Modelling of batch kinetics of aerobic carotenoid production using *Saccharomyces cerevisiae*. Biochemical Engineering Journal. 2016;**114**:226-236. DOI: 10.1016/j.bej.2016.07.004

[25] Yoo AY, Alnaeeli M, Park JK. Production control and characterization of antibacterial carotenoids from the yeast *Rhodotorula mucilaginosa* AY-01AH. Process Biochemistry. 2016;**51**:463-473. DOI: 10.1016/j.procbio.2016.01.008

[26] Dufossé L, Galaup P, Yaron A, Arad SM, Blanc P, Murthy KNC, Ravishankar G. Microorganisms and microalgae as sources of pigments for food use: A scientific oddity or an industrial reality? Trends in Food Science and Technology. 2005;**16**:389-406. DOI: 10.1016/j.tifs.2005.02.006

[27] Dufossé, L. Production of food grade pigments. Food Technology and Biotechnology. 2006;**44**:313-321.

[28] Kumar A, Vishwakarma HS, Singh J, Dwivedi S, Kumar M. Microbial pigments: Production and their applications in various industries. International Journal of Pharmaceutical, Chemical and Biological Sciences. 2015;**5**:203-212.

[29] Valduga E, Tatsch PO, Tiggemann HT, Toniazzo G, Zeni J, Luccio M. Produção de carotenoides: Microrganismos como fonte de pigmentos naturais. Química Nova. 2009;**32**:2429-2436. DOI: 10.1590/S0100-40422009000900036

[30] Ruiz-Sola MA, Rodríguez-Concepción M. Carotenoid biosynthesis in *Arabidopsis*: A colorful pathway. In: The *Arabidopsis* Book: American Society of Plant Biologists. 1st ed. 2012. 29. DOI: 10.1199/tab.0158

[31] Rodriguez-Amaya DB. Status of carotenoid analytical methods and in vitro assays for the assessment of food quality and health effects. Current Opinion in Food Science. 2015;**1**:56-63. DOI: 10.1016/j.cofs.2014.11.005

[32] Krinsky NI, Johnson E. Carotenoid actions and their relation to health and disease. Molecular Aspects of Medicine. 2005;**26**:459-516. DOI: 10.1016/j.mam.2005.10.001

[33] Woodside JV, McGrath AJ, Lyner N, McKinley MC. Carotenoids and health in older people. Maturitas. 2015;**80**:63-68. DOI: 10.1016/j.maturitas.2014.10.012

[34] WHO – World Health Organization. Ageing and Life Course [Internet]. 2012. Available from: http://www.who.int/world-health-day/2012/toolkit/background/en/ [Accessed: 2016-08-08].

[35] Mata-Gómez LC, Montañez JC, Méndez-Zavala A, Aguilar CN. Biotechnological production of carotenoids by yeasts: An overview. Microbial Cell Factories. 2014;**13**:12. DOI: 10.1186/1475-2859-13-12

[36] Rodrigues E, Mariutti LRB, Chisté RC, Mercadante AZ. Development of a novel micro-assay for evaluation of peroxyl radicalscavenger capacity: Application to carotenoids and structure-activity relationship. Food Chemistry. 2006;**135**:2103-2111. DOI: 10.1016/j.foodchem.2012.06.074

[37] Bartley GE, Scolnik PA. Plant carotenoids: Pigments for photoprotection, visual attraction and human health. The Plant Cell. 1995;**7**:1027-1038. DOI: 10.1105/tpc.7.7.1027

[38] Zoz L, Carvalho JC, Soccol VT, Casagrande CC, Cardoso L. Torularhodin and torulene: Bioproduction, properties and prospective applications in food and cosmetics – A review. Brazilian Archives of Biology and Technology. 2015;**58**:278-288. DOI: 10.1590/S1516-8913201400152

[39] Hill GE, Inouye CY, Montgomerie R. Dietary carotenoids predict plumage coloration in wild house finches. Proceeding of the Royal Society B: Biological Sciences. 2002;**269**:1119-1124. DOI: 10.1098/rspb.2002.1980

[40] Marova I, Carnecka M, Halienova A, Certik M, Dvorakova T, Haronikova A. Use of several waste substrates for carotenoid-rich yeast biomass production. Journal of Environmental Management. 2012;**95**:338-342. DOI: 10.1016/j.jenvman.2011.06.018

[41] Lin FY, Liu CI, Liu YL, Zhang Y, Wang K, Jeng WY, Ko TP, Cao R, Wang AH, Oldfield E. Mechanism of action and inhibition of dehydrosqualene synthase. Proceedings of the National Academy of Sciences. 2010;**107**:21337-21342. DOI: 10.1073/pnas.1010907107

[42] Farré G, Sanahuja G, Naqvi S, Bai C, Capell T, Zhu C, Christou P. Travel advice on the road to carotenoids in plants. Plant Science. 2010;**179**:28-48. DOI: 10.1016/j.plantsci.2010.03.009

[43] Papaioannou EH, Liakopoulou-Kyriakides M. Substrate contribution on carotenoids production in *Blakeslea trispora* cultivations. Food and Bioproducts Processing. 2010;**8**:305-311. DOI: 10.1016/j.fbp.2009.03.001

[44] Varzakakou M, Roukas T, Kotzekidou P. Mint: Effect of the ratio of (+) and (−) mating type of *Blakeslea trispora* on carotene production from cheese whey in submerged fermentation. World Journal of Microbiology Biotechnology. 2010;**26**:2151-2156. DOI: 10.1007/s11274-010-0398-3

[45] Khodaiyan F, Razavi SH, Mousavi SM. Optimization of canthaxanthin production by *Dietzia natronolimnaea* HS-1 from cheese whey using statistical experimental methods. Biochemical Engineering Journal. 2008;**40**:415-422. DOI: 10.1016/j.bej.2008.01.016

[46] Yang J, Tan H, Yang R, Sun X, Zhai H, Li K. Astaxanthin production by *Phaffia rhodozyma* fermentation of cassava residues substrate. Agricultural Engineering International. 2011;**13**:1-6.

[47] Razavi SH, Marc I. Effect of temperature and pH on the growth kinetics and carotenoid production by *Sporobolomyces ruberrimus* H110 using technical glycerol as carbon source. Iranian Journal of Chemistry and Chemical Engineering. 2006;**25**:59-64.

[48] Ip PF, Chen F. Production of astaxanthin by the green microalga *Chlorella zofingiensis* in the dark. Process Biochemistry. 2005;**40**:733-738. DOI: 10.1016/j.procbio.2004.01.039

[49] Abe K, Hattori H, Hirano M. Accumulation and antioxidant activity of secondary carotenoids in the aerial microalga *Coelastrella striolata* var. *multistriata*. Food Chemistry. 2007;**100**:656-661. DOI: 10.1016/j.foodchem.2005.10.026

[50] Vaquero I, Ruiz-Domínguez C, Márquez M, Vílchez C. Cu-mediated biomass productivity enhancement and lutein enrichment of the novel microalga *Coccomyxa onubensis*. Process Biochemistry. 2012;**47**:694-700. DOI: 10.1016/j.procbio.2012.01.016

[51] Harker M, Tsavalos A, Young AJ. Factors responsible for astaxanthin formation in the chlorophyte *Haematococcus pluvialis*. Bioresource Technology. 1996;**55**:207-217. DOI: 10.1016/0960-8524(95)00002-X

[52] Ip PF, Wong KH, Chen F. Enhanced production of astaxanthin by the green microalga *Chlorella zofingiensis* in mixotrophic culture. Process Biochemistry. 2004;**39**:1761-1766. DOI: 10.1016/j.procbio.2003.08.003

[53] Kleinegris DMM, Janssen M, Brandenburg WA, Wijffels RH. Continuous production of carotenoids from *Dunaliella salina*. Enzyme and Microbial Technology. 2011;**48**:253-259. DOI: 10.1016/j.enzmictec.2010.11.005

[54] Garcıa-Malea MC, Brindley C, Del Río E, Acien FG, Fernandez JM, Molina E. Modeling of growth and accumulation of carotenoids in *Haematococcus pluvialis* as a function of irradiance and nutrients supply. Biochemical Engineering Journal. 2005;**26**:107-114. DOI: 10.1016/j.bej.2005.04.007

[55] Del Campo JA, Moreno J, Rodrıguez H, Vargas MA, Rivas J, Guerrero MJ. Carotenoid content of chlorophycean microalgae: Factors determining lutein accumulation in *Muriellopsis* sp. (Chlorophyta). Journal of Biotechnology. 2000;**76**:51-59. DOI: 10.1016/S0168-1656(99)00178-9

[56] Hong ME, Choi SP, Park YI, Kim YK, Chang WS, Kim BW, Sim SJ. Astaxanthin production by a highly photosensitive *Haematococcus* mutant. Process Biochemistry. 2012;**47**:1972-1979. DOI: 10.1016/j.procbio.2012.07.007

[57] Hirschberg J, Harker M. Carotenoid-Producing Bacterial Species and Process for Production of Carotenoids Using Same. United States Patent 5,935,808. August 10, 1999.

[58] De Carvalho JC, Cardoso LC, Ghiggi V, Woiciechowski AL, de Souza Vandenberghe LP, Soccol CR. Microbial pigments. In: Brar SK, Dhillon GS, Soccol CR, editors. Biotransformation of Waste Biomass into High Value Biochemicals. Springer: New York; 2014. pp. 73-97. ISBN: 978-1-4614-8005-1

[59] Jaswir I, Noviendri D, Hasrini RF, Octavianti F. Carotenoids: Sources, medicinal properties and their application in food and nutraceutical industry. Journal of Medicinal Plants Research. 2011;**5**:7119-7131. DOI: 10.5897/JMPRx11.011

[60] Lin JH, Lee DJ, Chang JS. Lutein production from biomass: Marigold flowers versus microalgae. Bioresource Technology. 2015;**184**:421-428. DOI: 10.1016/j.biortech.2014.09.099

[61] Ravaghi M, Razavi SH, Mousavi SM, Sinico C, Fadda AM. Stabilization of natural canthaxanthin produced by *Dietzia natronolimnaea* HS-1 by encapsulation in niosomes. Food Science and Technology. 2016;**73**:498-504. DOI: 10.1016/j.lwt.2016.06.027

[62] Hernandez-Almanza A, Montañez J, Martínez G, Aguilar-Jimenez A, Contreras-Esquivel JC, Aguilar C N. Lycopene: Progress in microbial production. Trends in Food Science & Technology. 2016;**56**:142-148. DOI: 10.1016/j.tifs.2016.08.013

[63] Jing K, He S, Chen T, Lu Y, Ng I-S. Enhancing beta-carotene biosynthesis and gene transcriptional regulation in *Blakeslea trispora* with sodium acetate. Engineering Journal. 2016;**114**:10-17. DOI: 10.1016/j.bej.2016.06.015

[64] Panis G, Carreon JR. Commercial astaxanthin production derived by green alga *Haematococcus pluvialis*: A microalgae process model and a techno-economic assessment all through production line. Algal Research. 2016;**18**:175-190. DOI: 10.1016/j.algal.2016.06.007

[65] BBC Research. The Global Market for Carotenoids [Internet]. 2015. Available from: http://www.bccresearch.com/market-research/food-and-beverage/carotenoids-global-market-report-fod025e.html [Accessed: 2017-01-18]

[66] Spolaore P, Joaniss-Cassan C, Duran E, Isambert A. Commercial applications of microalgae. Journal of Bioscience and Bioengineering. 2006;**101**:87-96. DOI: 10.1263/jbb.101.87

[67] Raja R, Haemaiswarya S, Rengasamy R. Exploitation of *Dunaliella* for β-carotene production. Applied Microbiology and Biotechnology. 2007;**74**:517-523. DOI: 10.1007/s00253-006-0777-8d

[68] Singh RP. *Spirulina*: Health food for complete nutrition. Biotech Today. 2013;**3**:48-51. DOI: 10.5958/j.2322-0996.3.1.009

[69] Lian X, Liu L, Dong S, Wu H, Zhao J, Han Y. Two new monascus red pigments produced by Shandong Zhonghui Food Company in China. European Food Research and Technology. 2014;**240**:719-724. DOI: 10.1007/s00217-014-2376-8

[70] Galaup P, Sutthiwong N, Leclercq-Perlat MN, Valla A, Caro Y, Fouillaud M, Guérard F, Dufossé L. First isolation of *Brevibacterium* sp. pigments in the rind of an industrial red-smear-ripened soft cheese. Society of Dairy Technology. 2015;**68**:144-147. DOI: 10.1111/1471-0307.12211

[71] Kon IY, Gmoshinskaya MV, Safronova AI, Alarcon P, Vandenplas Y. Growth and tolerance assessment of a lutein-fortified infant formula. Journal Pediatric Gastroenterology Hepatology Nutrition. 2014;**17**:104-111. DOI: 10.5223/pghn.2014.17.2.104

[72] Prabhasankar P, Ganesan P, Bhaskar N, Hirose A, Stephen N, Gowda LR, Hosokawa M, Miyashita K. Edible Japanese seaweed, wakame (*Undaria pinnatifida*) as an ingredient in pasta: Chemical, functional and structural evaluation. Food Chemistry. 2009;**115**:501-508. DOI: 10.1016/j.foodchem.2008.12.047

[73] Kadam SU, Prabhasankar P. Marine foods as functional ingredients in bakery and pasta products. Food Research International. 2010;**43**:1975-1980. DOI: 10.1016/j.foodres.2010.06.007

[74] Niranjana R, Gayathri R, Stephen NM, Sugawara T, Hirata T, Myashita K, Ganesan P. Carotenoids modulate the hallmarks of cancer cells. Journal of Functional Foods. 2015;**18**:968-985. DOI: 10.1016/j.jff.2014.10.017

[75] Jomova K, Valko M. Health protective effects of carotenoids and their interactions with other biological antioxidants. European Journal of Medicinal Chemistry. 2013;**70**:102-110. DOI: 10.1016/j.ejmech.2013.09.054

[76] Shahidi F, Ambigaipalan P. Novel functional food ingredients from marine sources. Current Opinion in Food Science. 2015;**2**:123-129. DOI: 10.1016/j.cofs.2014.12.009

[77] Lai CS, Wu JC, Pan MH. Molecular mechanism on functional food bioactives for anti-obesity. Current Opinion in Food Science. 2015;**2**:9-13. DOI: 10.1016/j.cofs.2014.11.008

[78] Revuelta JL, Buey RM, Ledesma-Amaro R, Vandamme EJ. Microbial biotechnology for the synthesis of (pro) vitamins, biopigments and antioxidants: Challenges and opportunities. Microbial Biotechnology. 2016;**9**:564-567. DOI: 10.1111/1751-7915.12379

[79] Curhan SG, Stankovic KM, Eavey RD, Wang M, Stampfer MJ, Curhan GC. Carotenoids, vitamin A, vitamin C, vitamin E, and folate and risk of self-reported hearing loss in women. American Journal Clinical Nutrition. 2015;**102**:1167-1175. DOI: 10.3945/ajcn.115.109314

[80] Bakan E, Akbulut ZT, Inanç AL. Carotenoids in foods and their effects on human health. Akademik Gıda. 2014;**12**:61-68.

[81] Lu CC, Yen GC. Antioxidative and anti-inflammatory activity of functional foods. Current Opinion in Food Science. 2015;**2**:1-8. DOI: 10.1016/j.cofs.2014.11.002

[82] Castro-Puyana M, Pérez-Sánchez A, Valdés A, Ibrahim OHM, Suarez-Álvarez S, Ferragut JA, Micol V, Cifuentes A, Ilbáñez E, García-Cañas V. Pressurized liquid extraction of *Neochloris oleoabundans* for the recovery of bioactive carotenoids with anti-proliferative activity. Food Research International, in press. DOI: 10.1016/j.foodres.2016.05.021

[83] Zhang X, Zhao W, Hu L, Zhao L, Huang J. Carotenoids inhibit proliferation and regulate expression of peroxisome proliferators-activated receptor gamma (PPARc) in K562 cancer cells. Archives of Biochemistry and Biophysics. 2011;**512**:96-106. DOI: 10.1016/j.abb.2011.05.004

[84] Nishino H, Murakoshi M, Tokuda H, Satomi Y. Cancer prevention by carotenoids. Archives of Biochemistry and Biophysics. 2009;**483**:165-168. DOI: 10.1016/j.abb.2008.09.011

Chapter 9

Synthesis of Antioxidant Carotenoids in Microalgae in Response to Physiological Stress

Cecilia Faraloni and Giuseppe Torzillo

Additional information is available at the end of the chapter

http://dx.doi.org/10.5772/67843

Abstract

Carotenoids act as potential antioxidants, quenching energy of excited singlet oxygen and scavenging free radicals. Among microalgae, *Haematococcus*, *Chlamydomonas*, *Chlorella*, *Dunaliella* and diatoms and dinoflagellates, such as *Phaeodactylum* and *Isochrysis*, are able to synthesize large amount of carotenoids. The main function of carotenoids consists in absorbing light to perform photosynthesis, and some of them are constitutively present in the cells (primary carotenoids). The main primary carotenoids usually found are neoxanthin, violaxanthin, lutein, and β-carotene. To preserve cells from oxidative damage, their production may be increased, while other carotenoids may be synthesized *de novo*. In particular, under stress conditions such as high light exposure, nutrient starvation, change in oxygen partial pressure, and high or low temperatures, microalgal metabolism is altered and photosynthetic activity may be reduced. In these conditions, photosynthetic electrons transport is reduced, and the intracellular reduction level increase may be associated with the formation of free radicals and species containing singlet oxygen. In order to prevent damage from photooxidation, microalgae are able to adopt strategies to contrast these dangerous oxidant molecules. One of the most active mechanisms is to synthesize large amount of carotenoids, which can act as antioxidants.

Keywords: carotenoids, microalgae, antioxidant, stress

1. Introduction

Carotenoids are a class of natural lipid-soluble pigments mainly found in plants, algae, and photosynthetic bacteria. They play a central role in photosynthesis, both as light-harvesting complexes and as photoprotectors. However, it is generally believed that they function as passive photoprotectors (i.e., as a filter), reducing the amount of light that can reach the light-harvesting pigment complexes of photosystem II (PSII).

For their antioxidant properties, the role of carotenoids in human health has acquired importance in the recent years, mainly due to the attention toward the utilization of compounds obtained from natural sources.

Microalgae and cyanobacteria are photoautotrophic organisms that are exposed to high oxygen and radical stress in their natural environment, and consequently have developed several efficient protective systems against reactive oxygen species and free radicals [1]. They represent an almost untapped resource of natural antioxidants due to their enormous biodiversity, and the value of microalgae as a source of natural antioxidants is further enhanced by the relative ease of purification of target compounds [2].

Microalgae are capable, under stress conditions, of producing significant amounts of substances with high added value (antioxidant carotenoids, phenolic compounds, and polyunsaturated fatty acids), and for this reason, the study of the physiology of the growth of these microorganisms is of particular interest. In particular, carotenoids act by counteracting the effects of the damage caused by an excess of light and protecting the cells from oxidative damage.

Carotenoids are divided into two groups named primary and secondary carotenoids.

The primary carotenoids, such as the xanthophylls and β-carotene, are found in the chloroplast under standard conditions and are directly involved in performing photosynthesis for their role in the absorption of light energy. However, under stress conditions such as high light and nutrient deficiency, the provided energy may not be sustainable, and the content in primary carotenoids may increase, to dissipate the excess energy. Moreover, some photosynthetic microorganisms accumulate large amounts of secondary carotenoids in the cells, as a mechanism of photoprotection, in response to physiological stresses that induce the increase of reduction level inside the cells.

In particular, under high light stress conditions, the dissipation of the excess absorbed light energy occurs via the nonphotochemical quenching (NPQ) of chlorophyll fluorescence, a harmless nonradiative pathway of dissipation of energy. This defensive strategy involves the synthesis of antioxidant carotenoids, such as the secondary carotenoid astaxanthin, the pigment lutein, and the xanthophyll cycle pigments: violaxanthin, antheraxanhitn, and zeaxanthin [3–7]. Among the xanthophylls, also loroxanthin and fucoxanthin, mainly produced by marine strains such as *Phaeodactylum* and *Isochrysis*, have been found to be strong antioxidants.

Diatoms, such as *Phaeodactylum*, have a specific set of pigments with chlorophyll *c*, and they have an additive xanthophyll cycle, consisting in diadinoxanthin (Ddx), which can be deepoxidized to diatoxanthin (Ddx). These reactions lead to reduction of the singlet oxygen inside the cell, avoiding cellular damage. Among carotenoids, the ketocarotenoid astaxanthin has been shown to have a strong efficacy in quenching singlet oxygen.

Comparing the antioxidant activity of astaxanthin, β-carotene and the xanthophylls zeaxanthin and lutein with the one of alpha-tocopherol, a well-known noncarotenoid antioxidant, it is has been shown that these carotenoids are among the most powerful antioxidants [8].

Considering the role of carotenoids as quenchers of active oxygen species, they represent a very interesting natural source of antioxidant and antiaging substances.

Among photosynthetic microorganisms, the green unicellular microalga *Haematococcus pluvialis* is capable of producing a large amount of astaxanthin, a red pigment that starts to accumulate in the central part of the cell until the cell becomes entirely red. The other unicellular green microalga *Dunaliella salina* is well known for β-carotene production. In this microalga, the strong orange pigment is synthesized at one side of the cell, where it starts to accumulate in lipidic bodies, and then it continues to accumulate in the rest of the cell. Another big producer of antioxidant carotenoids is *Scenedesmus*, a colonial microalga able to produce large amounts of lutein, which makes the cells change their color from green to yellow.

Many studies on the physiology of microalgae have been carried out on the unicellular green alga *Chlamydomonas reinhardtii*. This microalga is considered a good model organism as it can be easily manipulated by means of genetic engineering; it has been the source of much information on photosynthetic responses to stress. Concerning the synthesis of carotenoids, particularly interesting were the studies on the xanthophyll cycle induction.

2. Physiology of the growth of microalgae

Photosynthetic microorganisms present a great variety of shape and size. Microalgae and cyanobacteria are distributed in a wide spectrum of habitat, having adapted their metabolism to complex and extreme environmental conditions (high salinity, extreme temperature, nutrient deficiency, and UV-radiation). To survive under such different harsh conditions, they have developed several strategies.

Each strain has its own optimal growth conditions, in regards to temperature, pH, salinity, light intensity, nutrient composition of the medium. Among these, one especially important parameter for photosynthetic microorganisms is light intensity.

The photosynthetic efficiency, i.e., transformation of light energy into chemical energy, is first and foremost limited by the fact that photosynthetic cells can only use light in the wavelength range from 400 to 700 nm so that only about 55% of incident solar light is useful to perform photosynthesis.

Moreover, it has to be considered that part of photosynthetic active radiation, about 10%, is reflected by the surface of the cells in the cultures; also, self-shading between cells further reduces the light utilization of each cell. Considering all these limitations, the percentage of light that can be used for photosynthesis is about 41%.

It is also important to consider some physiological limits of the photosynthetic apparatus, which makes it unable to utilize a light irradiation beyond a light intensity. Hence, about 20% of incident solar light is in excess, when it reaches the highest intensities in the central part of the day, and it is dissipated by heat and used to synthesize antioxidant pigments [9, 10].

In **Figure 1**, a typical light-curve response of *C. reinhardtii* is reported, comparing the electrons transport rate (ETR) of different strains with D1 protein mutation affecting photosynthetic performance with the wild type.

In this case, the photosynthetic activity is expressed as the capability to transfer electrons, but it could also be expressed as O_2 evolution, or CO_2 up-take. It is evident that different strains can have different behaviors at increasing light intensities, exhibiting different values of α, the slope of the first part of the curve, and different I_k value, i.e., the saturation irradiance, given as an intercept between α and ETR_{max}. According to the light saturation value, the strains can react differently, having different sensitivity to high light stress, and accumulating different levels of photooxidative stress.

For this reason, imposing a light stress inducing the carotenoids synthesis, as well as other stress conditions, such as nutrient limitation-starvation and excessive low or high temperature, is a useful approach in order to accumulate antioxidant compounds, but it is not convenient in terms of culture productivity, as under these limiting conditions, the growth is strongly affected.

One of the main physiological parameters used to monitor stress is the measurement of the photosynthetic activity, by evaluating oxygen evolution and Chla fluorescence measurement. In the presence of stress, the photosynthetic activity usually decreases, and it can be a useful indication on the kind of stress occurring to the cells. In particular, when the photosynthetic apparatus is impaired, light cannot be used efficiently, an accumulation of electrons on the electrons transport chain occurs and cells need to dissipate this excess of energy.

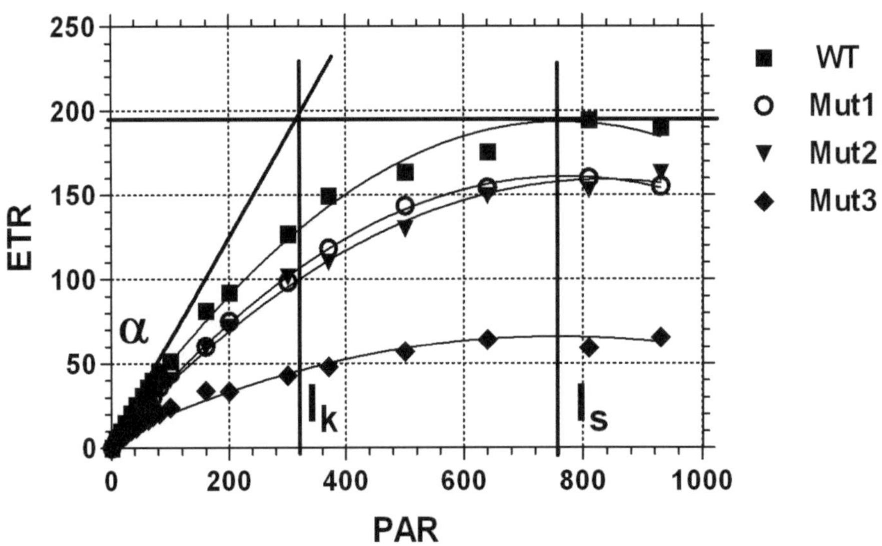

Figure 1. Comparison of different light induction curves in *Chlamydomonas reinhardtii* wild type (WT) and D1 protein mutant strains (mutation affecting the photosynthetic activity) Mut1, Mut2, and Mut3.

In response to this overreductive cellular environment condition, microalgae are able to produce a great variety of secondary metabolites, with antioxidant properties, which are biologically active and which cannot be found in other organisms [11, 12].

Among them, antioxidant compounds are the one to have attracted major interest for health and pharmaceutical industry, for their strong efficiency in preventing or delaying the damages caused by free radicals. Several synthetic antioxidants such as butylated hydroxyl anisole (BHA), butylated hydroxyl toluene (BHT), α-tocopherol, and propyl gallate have been used for limiting the oxidative damage, but they are strongly suspected to be responsible for a variety of side effects, such as liver damage and carcinogenesis. For this reason, a strong interest has been focused on finding natural products acting as antioxidants, safe, and effective.

3. Carotenoids: function and distribution in photosynthetic cells

The main functions of carotenoids consist in light absorption, to perform photosynthesis, and photoprotection to preserve the photosynthetic apparatus from photodamage. A role for carotenoids in cell differentiation, cell cycle regulation, growth factors regulation, stimulation of immune systems, intracellular signaling, and modulation of different kinds of receptors has been suggested [13].

However, for their antioxidant properties, they act as quenchers of active oxygen species and physiological stress, such as high light exposure, nutrient limitation or starvation, UV exposure, temperature fluctuation, anaerobiosis, and induce the metabolic pathways for the synthesis of these compounds.

These molecules are constituted by a C_{40} hydrocarbon backbone liable to structural modifications. According to their structure, carotenoids may be distributed in different ways into the cell compartments. In particular, they can be found within the inner section of the lipid bilayer of cell membranes, only if they are strict hydrocarbons like β-carotene or lycopene, or they can protrude into an aqueous environment from the membrane surface with a hydrophilic portion if they contain oxygen atoms, which confer them a more polar structure [14, 15]. Xanthophylls, such as lutein, fucoxanthin, neoxanthin, and xanthophyll cycle pigments, are among these more hydrophilic carotenoids. The presence of such carotenoids into the membranes may influence the thickness, fluidity, or permeability of them so that they can influence the stability of the cell membrane conferring it resistance, for instance, to ROS.

4. Photosynthetic and metabolic processes involved in the photoprotective responses in microalgae

Damage occurs when the free radical encounters another molecule and seeks to find another electron to pair with. The unpaired electron of a free radical pulls an electron off of a neighboring molecule, causing the affected molecule to behave like a free radical itself.

A range of biochemical and biophysical techniques had provided a good understanding of the events that occur during absorption of the light energy, triggering the primary and secondary electron transfer processes leading to water oxidation. These electron transport pathways involve the redox state of the component of the electron transport chain, the plastoquinone (PQ) pool, which has been widely investigated, for its implication in the regulation of photosynthetic processes.

Under oxidative stress conditions, there is an accumulation of reducing power inside the cells, which increases the reduction of PQ-pool. For this reason, the redox level of PQ-pool play a crucial role in the induction of physiological responses to stress, and it is important also for the synthesis of carotenoids.

It has been shown that there is an involvement of the redox state of PQ pool in the distribution of light energy during photosystem II (PS II) and photosystem I (PS I), i.e., state transitions. State 2 transition is promoted by the reduction of the PQ-pool and consists in the transfer of the light harvesting complex associated with PSII (LHCII) to the PSI, whereas under State 1 transition, which occurs when the PQ-pool is oxidized, the LHCII is associated with the PSII [16, 17]

The degree of reduction of PQ pool is related to a switch between linear and cyclic electron flow. With an over-reduced PQ pool (State 2), the PSI cross section increases and a cyclic electron transport is promoted, by contrast under oxidative conditions (State 1), the cross section of PSII is decreased and linear electron transport can be observed [18–20]. This is one of the strategies that photosynthetic cells employ to reduce the impact of strong light intensity on the photosynthetic apparatus, and it is triggered by the PQ-pool overreduction, and it is commonly associated with induction of carotenoid synthesis. Indeed, under these conditions, the acidification of the thylakoid lumen occurs, and this can activate some enzymes involved in the carotenogenesis. For instance, the deepoxidation of violaxanthin to zeaxanthin, via antheraxanthin, is promoted by low pH in the thylakoid lumen [5, 21, 22].

The synthesis of these carotenoids is important for the cells not only because the deepoxidation is a quenching reaction but also because xanthophylls have the ability to donate electrons [23] and act as inhibitors of the process of oxidation even at relatively small concentrations. Antioxidants also act as radical scavengers and convert radicals to less reactive species.

5. Stress-inducing the highest synthesis of antioxidant compounds

Which are the main kinds of stress to induce the carotenoids synthesis?

All those kinds of stress reducing growth and photosynthetic efficiency so that the excess of energy not used for growth (i.e., converted into biomass) is accumulated as reducing power and generates free radicals. Some of the well-known microalgae high producers of carotenoids are reported in **Table 1**. For each microalga the main stress factor inducing the carotenoids synthesis is reported with the respective antioxidant pigment. The detailed explanation is reported below in the text.

http://dx.doi.org/10.5772/67843

Microalgae	Carotenoids	Stress conditions
Haematococcus pluvialis	Asatxanthin; cantaxanthin; lutein	High light Nitrogen starvation
Dunaliella salina	β-Carotene	High light High temperature
Scenedesmus sp.	Lutein; β-carotene	High light Nutrient starvation
Phaeodactylum tricornutum	Diatoxanthin; fucoxanthin	High light Nutrient starvation
Isochrysis	Diatoxanthin; fucoxanthin	High light Nutrient starvation
Chlamydomonas reinhardtii	Zeaxanthin; lutein	High light Sulfur starvation Anaerobiosis

Table 1. Microalgae high producers of antioxidant carotenoids and stress conditions inducing their synthesis.

5.1. Light intensity

In particular, the exposure to high light is one of the typical stresses that microalgae may experience under environmental conditions. Indeed, during the central part of the day, the light irradiance may reach and exceed 1800 μmol photons m^{-2} s^{-1}.

A schematic explanation of the mechanism is reported in **Figure 2**.

Due to this accumulation of excess energy, leading to ROS formation, the synthesis of antioxidant carotenoids is induced in order to protect the cells from photodamage. Depending

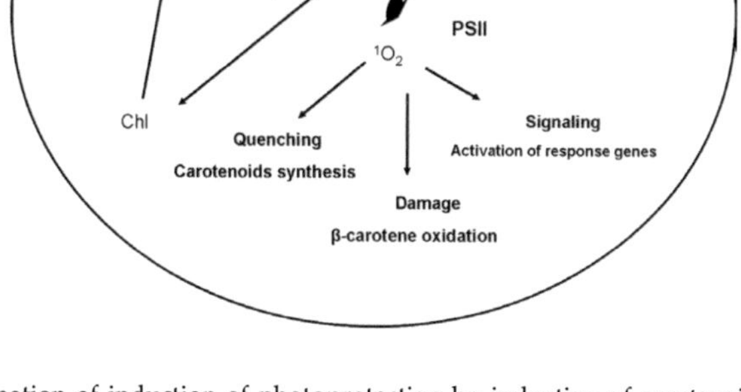

Figure 2. Schematic explanation of induction of photoprotection by induction of carotenoids synthesis by high light stress.

on the kind of light and on the strain, the mechanism of induction may follow different metabolic pathways. For instance, in case of sudden exposure to high light intensity, the cells may react with the induction of the xanthophyll cycle, which is known to occur very quickly, within 15–30 min [24]. This phenomenon has been widely reported in the microalga *C. reinhardtii,* which is considered a model organism for physiological and biochemical study on photosynthesis, because it can be easily manipulated for genetic study, and it can grow very easily both under photoheterotrophic and autotrophic conditions [25]. For this microalga, the induction of zeaxanthin synthesis has been detected within 10 minutes of exposure to 800 μmol photons m^{-2} s^{-1}, but a partial induction of violaxanthin de-epoxidation to antheraxanthin and then this one to zeaxanthin could be observed already at 300–350 μmol photons m^{-2} s^{-1} [26].

The induction of the xanthophyll cycle may affect also the synthesis of diatoxanthin by the de-epoxidation of diadinoxanthin, which represents an additional xanthophyll cycle in diatoms and dinoflagellates, such as *Phaeodactylum* and *Isochrysis,* respectively, among the main producers of this carotenoid. In *Phaeodactylum tricornutum,* a rapid diadinoxanthin to diatoxanthin conversion has been reported, within 15 min, during exposure to sunlight in outdoor cultures in tubular photobioreactors, with the highest diatoxanthin concentration reached in the central part of the day (highest light intensity) [27]. In addition, these microalgae are well known for the synthesis of fucoxanthin and important antioxidant carotenoid. Fucoxanthin is mainly naturally found in marine microalgae, associated with thylakoid membranes, and it works by transferring excitation energy to chlorophyll *a*, driving electrons to the electrons transport chain [28, 29]. Fucoxanthin is usually found to be 0.22–1.82% in the biomass of these microalgae, but it can reach much higher concentrations in *Isochrysis* cultured at proper light intensity, cell density, and mixing. In particular, it has been observed that in this microalga, the effect of self-shading and low light intensity induced an increase in total carotenoid concentration, probably due to the increase of photosystem number under low light, and consequently of the primary carotenoids.

Among the strongest antioxidant carotenoids, the pigment lutein can be overexpressed during high light exposure. It is a very interesting pigment, as it is constitutively present in most of photosynthetic cells, and its synthesis may increase under photooxidative stress. The microalga *Scenedesmus* produced high amounts of lutein (over 5 mg m^{-2} d^{-1}) in a tubular photobioreactor outdoor, under 1900 μmol photons m^{-2} s^{-1} and at 35°C [30]. In this case, the combined effect of high light and high temperature induced the increase of lutein. Indeed, usually, the optimal temperature of growth for microalgal strains is around 25–28°C.

Another carotenoid that usually increases during high light exposure is β-carotene. It is a pigment constitutively present in the microalgal cells, which may be oversynthesized under high light. One of the well-known microalgae for production of β-carotene is *D. salina* [31]. In laboratory conditions, it reached a production of 13.5 mg L^{-1} d^{-1} at light intensity in a range of 200–1200 μmol photons m^{-2} s^{-1}, at 30°C [32].

One of the most important secondary carotenoids produced by microalgae is the red pigment astaxanthin. It is a very powerful antioxidant primarily synthesized by *H. pluvialis,* mainly under high light. However, although its synthesis is not so rapid, as it takes 1 day of sunlight exposure to observe changes in the cells color, from green to red, it can reach a very high content, reaching 5% of the biomass. *H. pluvialis* has been widely studied for its astaxanthin production, due to its high productivity of this carotenoid, and for its robustness. Indeed, most of the studies carried out with *H. pluvialis* have been performed in outdoor cultures, using sunlight to induce astaxanthin production. These studies demonstrated that under environmental conditions, mainly in the summer period, and in very high illuminated areas, this microalga can grow and produce astaxanthin [33, 34].

5.2. Nutrient limitation

Nutrient limitation is another important stress condition inducing carotenoids synthesis and it is, like high light irradiance, a situation which can occur under environmental conditions. Macronutrient limitation, or starvation, is more incisive on the induction of protective responses than micronutrient limitation, as it directly affects growth, leading, mainly combined with light exposure, to the increase of reducing power, which is well known to activate defensive strategies such as the induction of the synthesis of certain carotenoids.

Nitrogen limitation is among the most studied nutrient-deprivation stress, as it is one of the most important elements in the cell, for its presence in proteins, enzymes, and because it is directly involved in the growth.

As previously reported in *Dunaliella* for β-carotene under high light stress, carotenoid increases in this microalga and this also occurs under nitrogen starvation. In particular, very interestingly, it has been shown that the increase in β-carotene content is concomitant with the synthesis of total fatty acid occurring under high light exposure and in combination with nitrogen starvation [35]. This can be explained by the fact that β-carotene is accumulated in lipid globules, in the cells, and it is supported by the findings that both lipid globules and β-carotene cannot be found when inhibitors of the fatty acid biosynthetic pathway are present [35]. At light intensity of 200 μmol photons m^{-2} s^{-1} under nitrogen starvation, a concentration of β-carotene of 2.7% of the biomass can be reached in *D. salina* [36].

A connection between lipid and carotenoid synthesis has been studied in *H. pluvialis*. In particular, the highest carotenoids accumulation has been observed with high light and nitrogen starvation combined, and under these conditions the astaxanthin content resulted more than two times higher than the control [37].

Under nitrogen starvation, astaxanthin synthesis is higher than in the control culture. Transition from the green stage to the red stage occurs during astaxanthin synthesis, due to the cytoplasmatic accumulation of the red pigment, which is observed within 20 h, reaching 1.4% of dry weight in the starved culture.

5.3. Overreduction of PQ-pool: anaerobiosis

Anaerobiosis is a condition that occurs when microalgal cells are cultivated in closed photobioreactors, in growth conditions that limit the photosynthetic activity; the oxygen evolution rate decreases reaching a value equal or lower than the oxygen respiration rate. Under light exposure, the electrons are driven by light, from water to the electrons transport chain, but if the photosynthetic apparatus is affected, it is not able to use the accumulated electrons, overreducing the cellular environment. Moreover, under anaerobic conditions, the respiration cannot eliminate these reducing electrons, for lack of oxygen that is the final electron acceptor, and therefore, the reduction level of PQ-pool cannot be dissipated.

It has been demonstrated that anaerobiosis has a strongly negative impact on the performance of photosynthetic cells, but on the other hand, it can be a useful means to activate certain metabolic processes sensitive to oxygen, for example, hydrogen production, in some microalgal strains like *Chlamydomonas reinhartdii* [38]. In this microalga, chlorophyll fluorescence and oxygen evolution measurements indicated a strong reduction of photosynthetic activity under sulphur starvation, which leads to the formation of a strongly reductive environment inside the cell compartments. This stress activates an antioxidative response promoting the synthesis of lutein and zeaxanthin [39]. Imposing anaerobic conditions to *C. reinhardtii* in complete medium, it was possible to observe a strong promotion of the xanthophyll cycle; however, under these conditions, the time of induction was not shorter than 5 h, contrary to the short time of induction at high light intensity. After this period, the zeaxanthin content was 12.63 mmol mol^{-1} Chl*a*. After 24 h it further increased, reaching 29.51 mmol mol^{-1} Chl*a*. Anaerobiosis induced the overexpression of all the xanthophyll pool, which increased by 15%, indicating a *de novo* synthesis of these xanthophylls, in particular violaxanthin, showing that this type of stress is not able to induce a rapid zeaxanthin synthesis but is strong enough to promote mechanisms of photoprotection on a longer time scale, with accumulation of large amounts of xanthophylls. In addition, increases in lutein content, which more than doubled, and of β-carotene, which increased by 90%, were observed. This strategy was able to preserve cells from photodamage. A very interesting aspect of the microalgal metabolism of carotenoids is that pigment composition may be adjusted by the cells according to the environmental conditions, and that some synthetic pathways can be very fast, in order to optimize the cellular performance and to save energy and storage [40]. In *C. reinahrdtii* cultures where the xanthophyll cycle had been induced, it has been shown that, after 1 h of aerobic dark adaptation, the pigments antheraxanthin and zeaxanthin decreased, as also did lutein and β-carotene, indicating the occurrence of a recovery. These findings underlined the very interesting peculiarity of microalgae, which consists in the strong capability to adapt to strong changes, in a different manner, according to the order of stress.

6. Importance of natural antioxidant compounds from microalgae and application in human health of antioxidants produced by microalgae

There is an increasing interest in the use of natural compounds in preventing and treating several diseases in humans, animals, and plants. For this reason, the research of a natural source of novel compounds with biological activity, in particular new and safe antioxidants, has gained a lot of importance.

http://dx.doi.org/10.5772/67843

Microalgae and cyanobacteria, under stress conditions, are capable of producing significant amounts of substances with high added value (antioxidant carotenoids, phenolic compounds, and polyunsaturated fatty acids), and for this reason, the study of the physiology of the growth of these microorganisms is of particular interest.

The secondary metabolites produced by photosynthetic organisms find numerous applications in the pharmaceutical, cosmetic, and food industries. In particular, the secondary carotenoids are widely used as antioxidants, acting as targets for highly reactive and toxic oxygen species, counteracting the effect of free radicals, and being effective as antiaging and anticancer agents.

Well known is the implication of carotenoids lutein and zeaxanthin in the pathologies of visual function, and the role of β-carotene in protecting the skin during exposure to the sun, and in the treatment of skin diseases.

It is well known that both lutein and zeaxanthin possess antioxidant properties due to their ability to quench singlet oxygen, reactive oxygen species, and free radicals [26, 41]. In particular, studies reported that an important role is played by lutein and zeaxanthin, constituents of the macular pigment, in the prevention of free radicals formation in the human retina, acting as quenchers [42–44]. This protective role against age-related macular degeneration makes these retinal carotenoids suitable for application as dietary supplements [45].

The antioxidant defense systems are important in maintaining good health, and therefore, an antioxidant-rich diet or antioxidant complements may be necessary as a health-protecting factor.

Interest in the employment of antioxidants from natural sources to increase the shelf life of food is considerably enhanced by the consumers' preference for natural ingredients and concerns about the toxic effects of synthetic antioxidants. Dietary antioxidants include three major groups: vitamins (vitamin C or ascorbic acid and vitamin E or tocopherols), phenols, and carotenoids, which are precursors of some vitamins.

7. Conclusions

Very interestingly, there is an interconversion among carotenoids, as some of them are precursor of others, and their metabolic pathways are often correlated. For example, in one case, the β-carotene can be the precursor of the xanthophyll violaxanthin. Particularly, under a strong oxidative stress, the induction of the xanthophyll cycle, with the deepoxidation of violaxanthin to zeaxanthin, via antheraxanthin, is concomitant to the decrease of β-carotene that contributes to the de novo synthesis of violaxanthin. This phenomenon has been reported in *C. reinhardtii*.

Moreover, zeaxanthin is reconverted to antheraxanthin and violaxanthin by the enzyme epoxidase. The plasticity of the carotenoid metabolism and the strong induction of their synthesis achievable in microalgae make this argument very interesting in terms of biotechnological applications.

Acknowledgements

This work was supported by Regione Toscana, Italy, in the framework of project PRAF and Officina Profumo Farmaceutica Santa Maria Novella, Florence, Italy, in the framework of a Contract Project.

Author details

Cecilia Faraloni* and Giuseppe Torzillo

*Address all correspondence to: faraloni@ise.cnr.it

Institute of Ecosystem Study, National Research Council, Italy

References

[1] Pulz O, Gross W: Valuable products from biotechnology of microalgae. Applied Microbiology and Biotechnology 2004;**65**(6):635–48.

[2] Li HB, Chen F, Zhang TY, Yang FQ, Xu GQ: Preparative isolation and purification of lutein from the microalgae *Chlorella vulgaris* by high-speed counter-current chromatography. Journal of Chromatography A 2001;**905**:151–155.

[3] Yamamoto HY: Biochemistry of the violaxanthin cycle in higher plants. Pure and Applied Chemistry 1979;**51**:6–648.

[4] Yamamoto HY, Nakayama TOM, and Chichester CO: Studies on the light and dark interconversion of leaf xanthophylls. Archives of Biochemistry 1962;**97**:168–173.

[5] Björkman O: High irradiance stress in higher plants and interaction with other stress factors. Photosynthesis Research 1987;**4**:11–18.

[6] Demmig-Adams B, Winter K, Kruger A, and Czygan FC: Light response of CO_2 assimilation, dissipation of excess excitation energy and zeaxanthin content of sun and shade leaves. Plant Physiology 1989;**90**:881–886.

[7] Eskling M, Arvidsson PO, Akerlund HE. The xanthophyll cycle, its regulation and components. Physiologia Plantarum 1997;**100**:806–816.

[8] Shimidzu N, Goto M, Miki W: Carotenoids as singlet oxygen quenchers in marine organisms. Fisheries Science 1996;**62**(1):134–137.

[9] Chaumont D, and Thepenier C: Carotenoid content in growing cells of *Haematococcus pluvialis* during sunlight cycle. Journal of Applied Phycology 1995;**7**:529–537.

[10] Masojidek J, Kopecky J, Koblizek, Torzillo G: The xanthophyll cycle in green algae (Chlorophyta): its role in the photosynthetic apparatus. Plant Biology 2004;**6**(3):342–349.

[11] Gonulol A, Ersanli E, Baytut O: Taxonomical and numerical comparison of epipelic algae from Balik and Uzun lagoon. Turkey. Journal of Environmental Biology 2009;**30**:777–784.

[12] Welker M, Dittmann E, von Dohren H: Cyanobacteria as source of natural products. Methods in Enzymology 2012;**517**:23–46.

[13] Fiedor J, and Burda K: Potential role of carotenoids as antioxidants in human health and disease. Nutrients 2014;**6**:466–488.

[14] Wísniewska A, and Subczynski WK: Effect of polar carotenoids in the shape of the hydrophobic barrier of phospholipids bilayers. Biochimica et Biophysica Acta 1998; **1368**:235–246.

[15] Wísniewska A, and Subczynski WK: Accumulation of macular xanthophylls in unsaturated membrane domains. Free Radical Biology & Medicine 2006;**40**:1820–1826.

[16] Gans P, and Rebeille F: Control in the dark of the plastoquinone redox state by mitochondrial activity in *Chlamydomonas reinhardtii*. Biochimica et Biophysica Acta 1990;**1015**:150–155.

[17] Allen JF: State transitions—a question of balance. Science 2003;**299**:1530–1532.

[18] Wollman FA: State transitions reveal the dynamics and flexibility of the photosynthetic apparatus. EMBO Journal 2001;**20**:3623–3630.

[19] Finazzi G, Rappaport F, Furia A, Fleischmann M, Rochaix JD, Zito F and Forti G: Involvement of state transitions in the switch between linear and cyclic electron flow in *Chlamydomonas reinhardtii*. EMBO Reports 2002;**3**(3):280–285.

[20] Albertsson PA: A quantitative model of the domain structure of the photosynthetic membrane. Trends in Plant Science 2001;**6**(8):349–354.

[21] Gilmore AM, and Yamamoto H: Linear models relating xanthophylls and acidity to non-photochemical fluorescence quenching. Evidence that antheraxanthin explains zeaxanthin-independent quenching. Photosynthesis Research 1993;**35**:67–68.

[22] Gilmore AM, Hazlett TL, and Govindjee: Xanthophyll cycle-dependent quenching of photosystem II chlorophyll *a* fluorescence-formation of a quenching complex with a short fluorescence lifetime. Proceedings of the National Academy of Science USA 1995;**92**:2273–2277.

[23] Mandal S, Yadav S, Yadav S, Nema RK: Antioxidants a review. Journal of Chemical and Pharmaceutical Research 2009;**1**:102–104.

[24] Niyogi KK, Bjorkman O, Grossman AR: *Chlamydomonas* xanthophyll cycle mutants identified by video imaging of chlorophyll fluorescence quenching. The Plant Cell 1997; **9**:1369–1380.

[25] Heifetz PB, Förster B, Osmond CB, Giles LJ, Boynton JE: Effects of acetate on facultative autotrophy in *Chlamydomonas reinhardtii* assessed by photosynthetic measurements and stable isotope analyses. Plant Physiology 2000;**122**(4):14–1446.

[26] Niyogi KK, Björkman O, and Grossman A: The role of specific xanthophylls in photoprotection. Proceedings of the National Academy of Sciences 1997;**94**:14162–14167.

[27] Torzillo G, Faraloni C, Silva AM, Kopecky J, Pilný J, Masojídek J: Photoacclimation of *Phaeodactylum tricornutum* (Bacillariophyceae) cultures grown in outdoors photobioreactors and open ponds. European Journal of Phycology 2012;**47**(2):169–181.

[28] Jin E, Polle JEW, Lee HK, Hyun SM, Chang M: Xanthophylls in microalgae from biosynthesis to biotechnological mass production and application. Journal of Microbiology and Biotechnology 2003;**13**:165–174.

[29] Mulders KJM, Lamers PP, Martens DE, Wijffels RH: Phototrophic pigment production with microalgae: biological constraints and opportunities. Journal of Phycology 2014;**50**:229–242.

[30] Fernández-Sevilla JM, Acién-Fernández FG, Molina-Grima E: Biotechnological production of lutein and its applications. Applied Microbiology and Biotechnology 2010; **86**:27–40.

[31] García-González M, Moreno J, Manzano JC, Florêncio FJ, Guerrero MG: Production of *Dunaliella salina* biomass rich in 9-cis-β-carotene and lutein in a closed tubular photobioreactor. Journal of Biotechnology 2005;**115**:81–90.

[32] Kleinegris DMM, Janssen M, Brandeburg WA, Wijffels RH: Continuous production of carotenoids from *Dunaliella salina*. Enzyme Microbiology and Biotechnology 2011;**85**:289–295.

[33] Recht L, Töpfer N, Batushansky A, Sikron N, Zarka A, Gibon Y, Nikolosky Z, Fait A, Boussiba S: Metabolite profiling and integrative modeling reveal metabolic constraints for carbon portioning under nitrogen starvation in the green algae *Haematococcus pluvialis*. JBC 2014;289(44):30387–403.

[34] Torzillo G, Goksan T, Faraloni C, Kopecky J and Masojidek J: Interplay between photochemical activities and pigment composition in an outdoor culture of *Haematococcus pluvialis* during the shift from the green to red stage. Journal of Applied Phycology 2003;**15**:127–136.

[35] Rabbani S, Beyer P, von Lintig J, Hugueney P, Kleinig H: Induced β-carotene synthesis driven by triacylglycerol deposition in the unicellular alga *Dunaliella bardawil*. Plant Physiology 1998;**116**:12–1248.

[36] Lamers PP, Janssen M, De Vos RCH, Bino RJ, Wijffels RH: Carotenoids and fatty acid metabolism in nitrogen starved *Dunaliella salina*, a unicellular green microalgae. Journal of Biotechnology 2012;**162**:21–27.

[37] Liang C, Zhai Y, Xu D, Ye N, Zhang X, Wang Y, Zhang W, Yu J: Correlation between lipid and carotenoid synthesis and photosynthetic capacity in *Haematococcus pluvialis*. Gracias Y Aceites 2015;66(2):e077.

[38] Melis A, Zhang L, Forestier M, Ghirardi ML, Seibert M: Sustained photobiological hydrogen gas production upon reversible inactivation of oxygen evolution in the green alga *Chlamydomonas reinhardtii*. Plant Physiology 2000;**122**:127–136.

[39] Faraloni C, Torzillo G: Xanthophyll cycle induction by anaerobic conditions under low light in *Chlamydomonas reinhardtii*. Journal of Applied Phycology 2013;**25**:1457–1471.

[40] Demmig-Adams B: Carotenoids and photoprotection in plants: a role for the xanthophyll zeaxanthin. Biochimica et Biophysica Acta 1990;**1020**:1–24.

[41] Govindjee, and Seufferheld MJ: Non-photochemical quenching of chlorophyll a fluorescence: early history and characterization of two xanthophylls-cycle mutants of *Chlamydomonas reinhardtii*. Functional Plant Biology 2002;**29**:1141–1155.

[42] Bhosale P, and Bernstein PS: Microbial xanthophylls. Applied Microbiology and Biotechnology 2005;**68**:445–455.

[43] Lornejad-Schäfer MR, Lambert C, Breithaupt DE, Biesalsky HK, and Frank J. TITLE. European Journal of Nutrition. 2007;**46**(2):79–86.

[44] Stringham JM, Hammond BR: Macular pigment and visual performance under glare conditions. Optometry and Vision Science 2008;**85**(2):82–88.

[45] Mozaffarieh M, Sacu S, Wedrich A: The role of the carotenoids, lutein and zeaxanthin, in protecting against age-related macular degeneration: a review based on controversial evidence. Nutritional Journal 2003;2:20.

Chapter 10

Carotenoid Production by *Corynebacterium*: The Workhorse of Industrial Amino Acid Production as Host for Production of a Broad Spectrum of C40 and C50 Carotenoids

Nadja A. Henke, Petra Peters-Wendisch and
Volker F. Wendisch

Additional information is available at the end of the chapter

http://dx.doi.org/10.5772/67631

Abstract

Corynebacterium glutamicum is used as a workhorse of industrial biotechnology for more than 60 years since its discovery as a natural glutamate producer in the 1950s. Nowadays, L-glutamate and L-lysine are being produced with this GRAS organism in the million-ton scale every year for the food and feed markets, respectively. Sequencing of the genome and establishment of a genetic toolbox boosted metabolic engineering of this host for a broad range of industrially relevant compounds ranging from bulk chemicals to high-value products. Carotenoids, the colourful representatives of terpenoids, are high-value compounds whose bio-based production is on the rise. Since *C. glutamicum* is a natural producer of the rare C50 carotenoid decaprenoxanthin, this organism is well suited to establish terpenoid-overproducing platform strains with the help of metabolic engineering strategies. In this work, the carotenogenic background of *C. glutamicum* and the metabolic engineering strategies for the generation of carotenoid-overproducing strains are depicted.

Keywords: *Corynebacterium*, C40/C50 carotenoids, biotechnological production, metabolic engineering, decaprenoxanthin, β-carotene, astaxanthin

1. Introduction

Carotenoids are the dominant pigments for the colouration of food, feed and beverages. The annual demand of the feed additive astaxanthin, for example, is estimated to be 130 tons for aquaculture and poultry breeding [1]. Besides their yellow-to-red colouring properties, this group of terpenoids is drawing attention in the healthcare industry due to their high antioxidant activities. Since the demand of naturally produced carotenoids is rising, and the fact that extraction of these high-value compounds from plant material is rather cost-inefficient, alternative and flexible production systems are favoured [2].

Corynebacterium glutamicum is a workhorse for the million-ton-scale production of L-glutamate and L-lysine. As a natural carotenoid producer, this biotechnological established microorganism is in focus as a suitable cell factory for natural carotenoid production. The biotechnological application of this genetically accessible bacterium has been exploited for the production of various natural and non-native products [3], including its potential to produce a range of industrially relevant carotenoids.

2. Carotenoid production with *Corynebacterium glutamicum*

2.1. *Corynebacterium glutamicum* as an established cell factory

For more than 60 years, *C. glutamicum* is used as a cell factory for the production of amino acids. Because of its GRAS status, this microbe has a long history in the food and feed industry in the million-ton scale [4]. Since its discovery as a natural glutamate producer in the 1950s [4], its production spectrum was broadened. Lysine is the second biggest production volume being generated by this microbe [5, 6]. Sequencing of its genome [7], the development of a well-filled genetic toolbox [8] and establishment of large-scale fermentations [9] boosted investigations of a wide set of industrially relevant compounds.

This cell factory can naturally utilise glucose, fructose, sucrose, mannitol, arabitol, propionate and acetate under aerobic conditions [10, 11]. Moreover, metabolic engineering enabled utilisation of alternative carbon and energy sources such as glycerol [12], amino sugars [13, 14], β-glucans [15], levoglucosan [16], pentoses [17] and starch [18]. Thus, for industrial fermentations usage of carbon sources which are available in high quantities at low prices is possible, while competition with food and feed resources can be avoided.

Although *C. glutamicum* was discovered as a natural glutamate producer, nowadays, several metabolically engineered production strains are available. Besides the proteinogenic amino acids, also non-proteinogenic amino acid like gamma-aminobutyrate (GABA) [19, 20] and citrulline [21], diamines [22, 23], alcohols [24, 25] and organic acids [26, 27] have been produced. Moreover also high-value compounds including the sesquiterpenoid valencene [28] and the C40 carotenoids β-carotene and astaxanthin [29, 30] can be synthesised with this microbe.

The broad product spectrum from bulk compounds, building blocks, food and feed additives and pharmaceutical and bioactive compounds indicates that *C. glutamicum* has developed to a chassis organism of metabolic engineering. Therefore, several efforts for systematic reduction and optimisation of the *C. glutamicum* genome have been made aiming on a reduced metabolic complexity and strength for future purposes of genetic engineering of new routes for new products [31, 32].

2.2. Carotenogenesis in *Corynebacterium glutamicum*

C. glutamicum is a yellow-pigmented soil bacterium due to the accumulation of a rare C50 carotenoid decaprenoxanthin and its glucosides [33]. Long-chain C50 carotenoids have been mainly isolated from extremely halophilic archaea [34, 35] and Gram-positive bacteria of the order *Actinomycetales* [36]. Among these species few other corynebacteria are known to produce carotenoid pigments such as *C. michiganense* [37], *C. erythrogenes* [38], *C. fascians* [39] and *C. poinsettia* [40]. Nevertheless, the respective pathways are yet poorly understood due to a lack of complete genome information. Carotenogenesis in *C. glutamicum* has been functionally characterized. The genome of *C. glutamicum* possesses two *crt* operons [7], whereas one of them encodes for all the necessary enzymes responsible for the conversion of the C5 precursor molecules dimethylallyl pyrophosphate (DMAPP) and isopentenyl pyrophosphate (IPP) to the C50 carotenoid decaprenoxanthin [41].

2.2.1. *Organisation of carotenogenic genes within the chromosome*

The precursor molecules derive from the methylerythritol phosphate (MEP) pathway, whose respective genes are partially clustered within the chromosome (**Figure 1**) [30]. The genes *dxs* (cg2083), *ispH* (cg1164) and *idi* (cg2531) are monocistronic, while *dxr* (cg2208), *ispD* (cg2945), *ispE* (cg1039), *ispF* (cg2944) and *ispG* (cg2206) are organised in putative operons. *IspE* is the third gene of the operon cg1037-*ksgA*-*ispE*-cg1040-*pdxK* with genes for a putative resuscitation-promoting factor (cg1037), putative dimethyl adenosine transferase (KsgA) and putative pyridoxamine kinase (PdxK). The genes *ispD* and *ispF* are encoded in the cg2946-*ispDF* operon with cg2946, which codes for a putative CarD-like transcriptional regulator. The genes *dxr* and *ispG* are organised in a transcriptional unit separated by an uncharacterised gene (cg2207) putatively encoding a membrane-embedded Zn-dependent protease. In *C. glutamicum* two prenyltransferases have been characterised [42] which both yield GGPP. Interestingly, the major prenyltransferase IdsA is not encoded within the major *crt* operon [42] but annotated in an operon that is also containing $\alpha(1\rightarrow6)$ mannopyranosyltransferase (**Figure 1**) [43]. CrtE is the second functional GGPP synthase, which is the first gene of the major *crt* operon (**Figure 1**). Both *crt* operons contain a *crtB* gene; however, the one from the small *crt* operon is weakly expressed [41]. Although both *crt* operons contain gene coding for a phytoene desaturase, only CrtI from the major operon is functional [41]. In addition, *crtEb*, $crtY_e$ and $crtY_f$ are part of the major *crt* operon which is necessary for decaprenoxanthin formation (**Figure 4**). Glucosyltransferase gene *crtX* is located in proximity of the major *crt* operon.

Although little is known about carotenogenesis in other corynebacteria, the genomic organisation of corresponding *crt* genes seems to be conserved [41].

2.2.2. *Biosynthesis of decaprenoxanthin*

The terpenoid precursor molecules DMAPP and IPP derive from the MEP pathway that uses pyruvate and GAP as substrates from central metabolism in *C. glutamicum*. Eight genes are encoding the enzymes which are necessary to build up the C5 precursors under consumption of NADPH, ATP and CTP (**Figure 2**). It has to be mentioned that DXP, the first intermediate

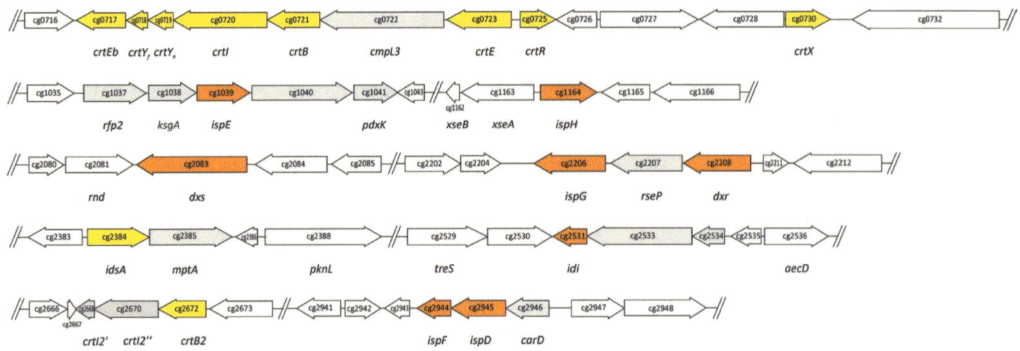

Figure 1. Genomic organisation of carotenogenic genes of *Corynebacterium glutamicum*. The MEP pathway genes involved in the synthesis of the isoprenoid precursors IPP and DMAPP from pyruvate and GAP are depicted in yellow; genes encoding for the decaprenoxanthin biosynthesis pathway from IPP and DMAPP are depicted in orange. Genes that are organised in operons with the respective MEP pathway or decaprenoxanthin pathway genes are shown in grey. Gene names, IDs and respective protein products: *crtEb* (cg0717), lycopene elongase; *crtY*$_f$ (cg0718) and *crtY*$_e$ (cg0719), hetero-dimeric C50 ε-cyclase; *crtI* (cg0720), phytoene desaturase; *crtB* (cg0721), phytoene synthase; *cmpL3* (cg0722) RND transporter, corynebacterial membrane protein [44]; *crtE* (cg0723), GGPP synthase; *crtR* (cg0725), transcriptional regulator of carotenoid biosynthesis; *crtX* (cg0730), carotenoid glycosyltransferase; *rfp2* (cg1037), resuscitation-promoting factor 2 [45]; *ksgA* (cg1038), putative 16S ribosomal RNA methyltransferase; *ispE* (cg1039), 4-diphosphocytidyl-2-C-methyl-D-erythritol kinase; *pdxK* (cg1041), putative pyridoxamine kinase; *xseB* (cg1162), putative exodeoxyribonuclease VII small subunit; *xseA* (cg1163), putative exodeoxyribonuclease VII large subunit; *ispH* (cg1164), 4-hydroxy-3-methylbut-2-en-1-yl diphosphate reductase; *rnd* (cg2081), putative ribonuclease D; *dxs* (cg2083), 1-deoxy-D-xylulose-5-phosphate synthase; *ispG* (cg2206), (E)-4-hydroxy-3-methylbut-2-enyl-diphosphate synthase; *rseP* (cg2207), putative membrane-embedded Zn-dependent protease; *dxr* (cg2208), 1-deoxy-D-xylulose-5-phosphate reductoisomerase; *idsA* (cg2384), GGPP synthase; *mptA* (cg2385), α(1→6) mannopyranosyltransferase [43]; *treS* (cg2529), trehalose synthase [46]; *idi* (cg2531), isopentenyl-diphosphate δ-isomerase; *aecD* (cg2536), cystathionine β-lyase [47]; *pknL* (cg2388), serine/threonine protein kinase [48]; *ispF* (cg2944), 2-C-methyl-D-erythritol 2,4-cyclodiphosphate synthase; *ispD* (cg2945), 2-C-methyl-D-erythritol 4-phosphate cytidylyltransferase and *carD* (cg2946), putative CarD-like transcriptional regulator.

of the MEP pathway, is also a precursor molecule for thiamine biosynthesis in *C. glutamicum*. Two prenyltransferases IdsA and CrtE use one molecule DMAPP and three molecules IPP to form GGPP (C20) (**Figure 2**) [41]. Two GGPP molecules can be condensed by a phytoene synthase CrtB to phytoene, a colourless C40 structure (**Figure 2**). The red lycopene is formed via the phytoene desaturase CrtI [17].

Lycopene is a central metabolite for all C40 and C50 carotenoids (**Figures 3** and **4**). In *C. glutamicum* this C40 carotenoid is further elongated by the lycopene elongase CrtEb with two DMAPP molecules to form the linear C50 carotenoid flavuxanthin. Finally, a heterodimeric ε-cyclase CrtY$_e$Y$_f$ introduces ε-cyclic moieties [49] in that linear C50 structure to form decaprenoxanthin (**Figure 4**). Decaprenoxanthin can be glycosylated either at one or at both terminal hydroxy groups by CrtX [50]. The physiological function remains unknown; however, glycosylation of end groups generally yields a more polar molecule structure in comparison to the free decaprenoxanthin and might change integrity into the membrane.

2.2.3. The unusual cell envelope of C. glutamicum *and carotenoid accumulation*

The cell envelope of the Gram-positive bacterium *C. glutamicum* possesses a special and complex cell wall structure [51, 52]. Besides the inner plasma membrane, the peptidoglycan and

http://dx.doi.org/10.5772/67631

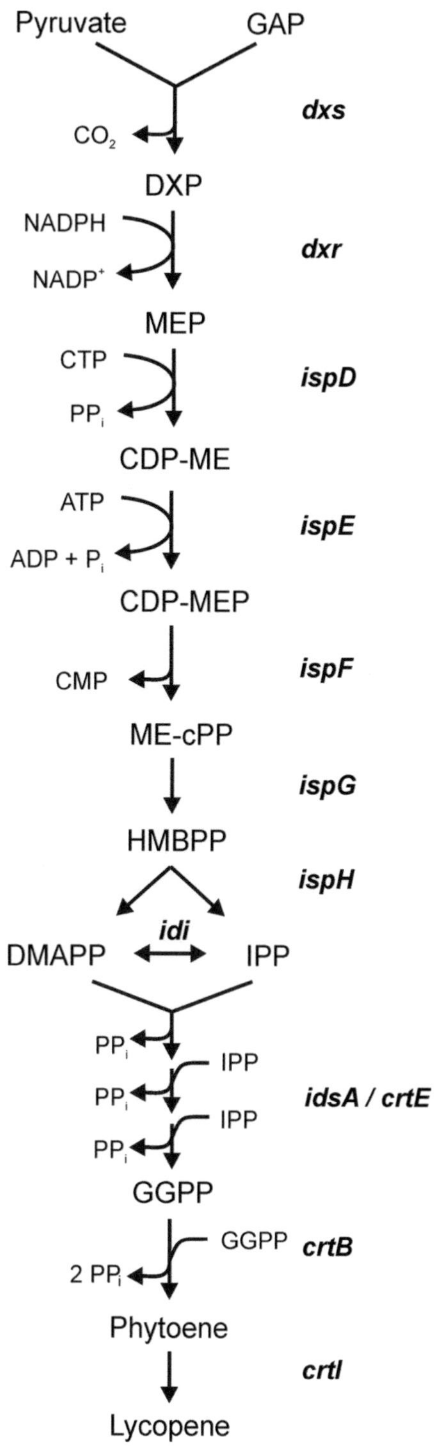

Figure 2. Metabolic pathway of lycopene production in *Corynebacterium glutamicum* starting from GAP and pyruvate. Genes are shown next to the reaction catalysed by the encoded enzyme (see **Figure 1**; *crtE*/*idsA*: prenyltransferase; *crtB*: phytoene synthase and *crtI*: phytoene desaturase).

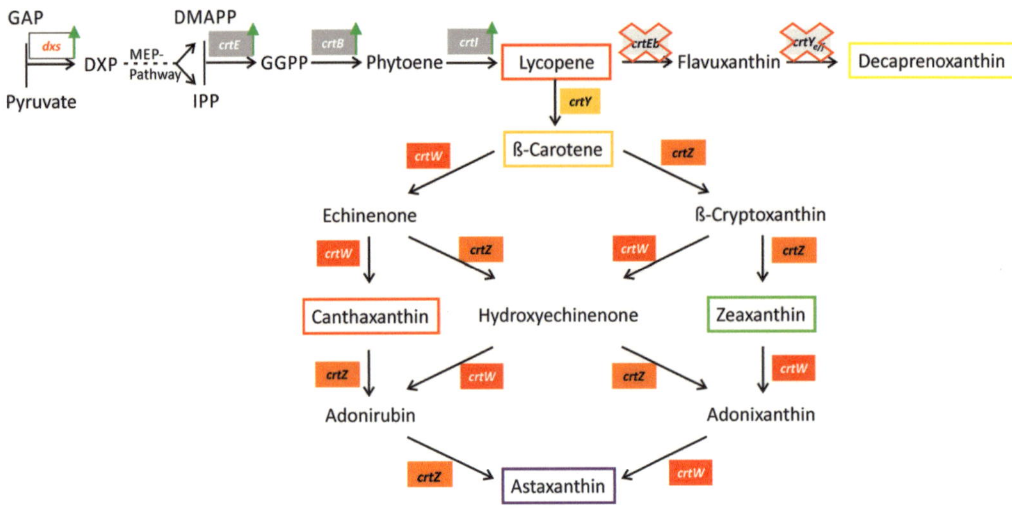

Figure 3. Metabolically engineered platform strains for production of cyclic C40 carotenoids. The biosynthesis of C40 carotenoids from the precursor molecules dimethylallyl pyrophosphate (DMAPP) and isopentenyl pyrophosphate (IPP) is depicted. Genes are shown next to the reaction catalysed by the encoded enzyme (*crtE*: GGPP synthase; *crtB*: phytoene synthase; *crtI*: phytoene desaturase; *crtEb*: lycopene elongase; *crtYe/f*: C45/50 carotenoid ε-cyclase; *crtY*: lycopene cyclase; *crtZ*: β-carotene hydroxylase and *crtW*: β-carotene ketolase). Endogenous genes are highlighted in grey boxes and overexpressions indicated by green arrows. Heterologous genes are highlighted in coloured boxes.

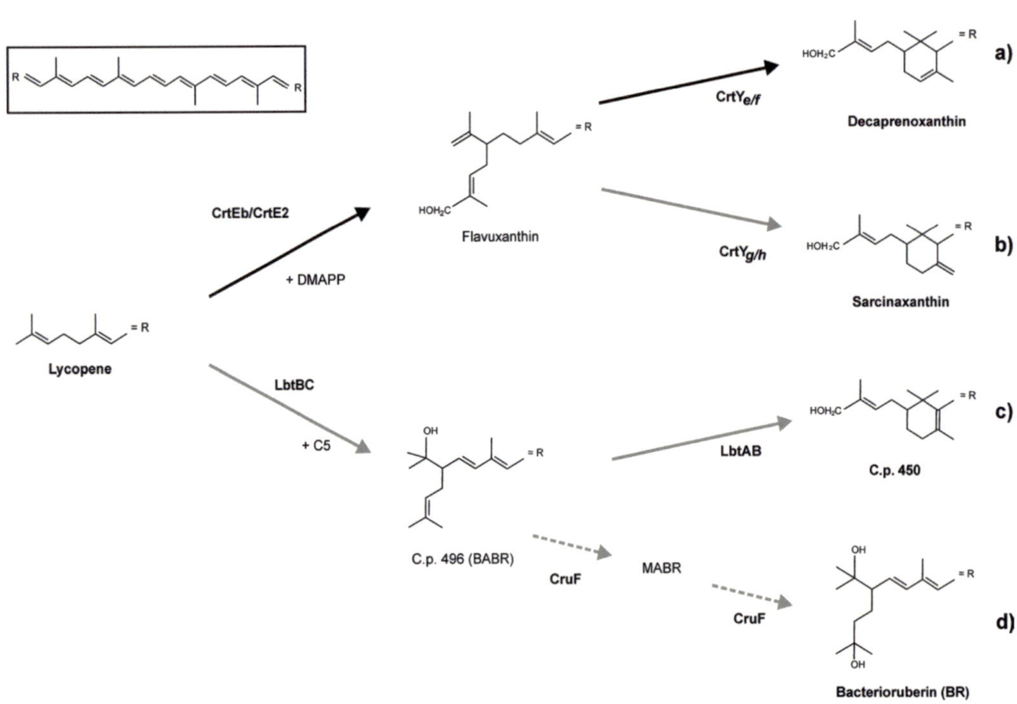

Figure 4. Biosynthesis of C50 carotenoids starting from the precursor lycopene. The native pathway of decaprenoxanthin biosynthesis (a) is shown with black arrows. Established non-native biosynthesis pathways of sarcinaxanthin (b) and C.p. 450 (c) are indicated by grey arrows. Possible biosynthesis of bacterioruberin (d) from C.p. 496 (BABR) is indicated by dashed grey arrows.

arabinogalactan layer forms a polysaccharide barrier that is esterified to mycolic acids. The presence of the mycolic acid layer is a phylogenetic trait of *Corynebacterianeae* [52]. The plasma membrane consists of polar lipids, mainly phospholipids which form together with proteins the typical lipid bilayer.

Carotenoids are usually attached to or span membranes due to their lipophilic character and rigid structure. Association of carotenoids to membranes often results in a decreased water permeability and increased firmness, thus supporting membrane stability [53, 54]. It is hypothesized that this is closely linked to their function supporting resistances to osmotic stresses, heat or radiation [54–56]. Moreover, it was shown that incorporation of carotenoids into a membrane is more efficient when the carbon backbone length of the carotenoid correlates with the thickness of the phospholipid bilayer [57]. Although decaprenoxanthin is a C50 carotenoid, it is assumed to be integrated into the plasma membrane [58, 59] as it was also shown for most C40 carotenoids of other bacteria [53, 56, 60].

2.3. Metabolic engineering of *Corynebacterium glutamicum* for carotenoid production

Since *C. glutamicum* naturally produces a rare cyclic C50 carotenoid, its potential to produce industrially relevant C40 carotenoids was elucidated over the last years in more detail. First, it was shown that production of both non-native C50 and C40 carotenoids was possible with this production host [50, 61]. Secondly, improvement of the MEP pathway yielded enhanced production [30]. Finally, the production of industrially relevant cyclic C40 carotenoids was shown on the basis of balancing of the enzyme quantities and on the basis of a screen for suitable enzymes for enhanced production [29].

2.3.1. Design of a platform strain for the production of the central intermediate lycopene

For production of non-native C40 and C50 carotenoids, endogenous decaprenoxanthin production has to be avoided (**Figure 3**). For this reason, a prophage-cured *C. glutamicum* strain MB001 [62] was metabolically engineered by deleting the genes *crtEb* and $crtY_eY_f$, resulting in the biosynthesis of the central intermediate lycopene [41, 50]. The supply of the precursor molecule DMAPP and its isomer IPP was successfully engineered through improved expression of *dxs*, encoding for the first committed step in the MEP pathway, on the basis of a chromosomal promoter exchange [30]. Furthermore, it was shown that overproduction of the prenyltransferase (CrtE), phytoene synthase (CrtB) and phytoene desaturase (CrtI) strongly enhanced lycopene production [50]. Chromosomal integration of the artificial operon *crtEBI* under the control of a strong constitutive *tuf* promoter in a $\Delta crtEbY_eY_f$ strain yielded a red phenotype [29].

2.3.2. Metabolic engineering for cyclic C40 carotenoid productions

On the basis of lycopene-producing platform strains, a collection of cyclic C40 carotenoids could be generated via different metabolic engineering strategies. First of all, β-cyclisation of lycopene was accomplished via heterologous overexpression of *crtY* from *Pantoea ananatis* [29, 30]. Here, either plasmid-driven or genome-based expressions revealed complete conversion of lycopene to β-carotene (**Figure 3**). This might rely on the fact that β-carotene has a more polar structure due to its β-ionone rings and thus integrates more efficiently in the phospholipid bilayer than lycopene. Heterologous expression of β-carotene hydroxylase from *P. ananatis* in such a β-carotene–accumulating strain yielded production of zeaxanthin [30]. Additional plasmid-driven overproduction of a β-carotene ketolase from *Brevundimonas aurantiaca* showed accumulation of the red C40 carotenoid astaxanthin for the first time in *C. glutamicum* [30]. Analysis of the carotenoid profile showed that conversion of β-carotene to its oxygenated derivatives was not efficient maybe due to an unbalanced expression of involved genes.

Pathway balancing was performed through balancing of the enzyme levels of β-carotene ketolase and hydroxylase (**Figure 3**). In a combinatorial approach, enzyme quantities were varied on the basis of varied translation initiation rates [29]. The corresponding genes were assembled in an expression vector under the strong constitutive *tuf* promoter using different ribosome-binding sites, different spacing lengths and different start codons. The translation initiation rate is depending on the free binding energy of the ribosome-binding site and the 16S rRNA as well as on the free energy of secondary structures of the mRNA [29]. Secondary structures between 5′UTR and coding region are depending on the expressed gene and were calculated with the RBS calculator [63]. Analysis revealed that the higher the TIRs of both genes, the higher the astaxanthin titer [29] in recombinant *C. glutamicum*. With this approach an astaxanthin content of 0.3 mg/g CDW was accomplished indicating that aside pathway balancing also enzyme activities or stabilities could be limiting the production.

The analysis of carotenogenesis identified a transcriptional regulator CrtR [64]. This MarR regulator is binding to the promoter sequence of *crtE* and thus inhibits transcription of the carotenogenic genes. Deletion of the corresponding gene located divergent to the major *crt* operon enhanced both native and non-native carotenoid productions. The production titers of non-native C50 carotenoids sarcinaxanthin and C.p. 450 improved 1.5 and 2-fold, respectively. For the deregulated β-carotene–accumulating strain production, titer doubled to approximately 12 mg/g CDW [29]. This platform strain was used for a second approach to produce astaxanthin with recombinant *C. glutamicum*. In the deregulated β-carotene–producing platform strain, a screening of potential β-carotene ketolases and hydroxylases of rather uncharacterised gene donors for production of canthaxanthin and zeaxanthin was done. Two enzymes from *Fulvimarina pelagi* showed the best conversion of β-carotene to the oxygenated derivatives [29]. Moreover, a combination of those two genes with a two-vector system resulted in a *C. glutamicum* strain producing 1.6 mg/g CDW of astaxanthin [29], which is the highest reported titer obtained with this organism. Since *C. glutamicum* is fast growing even in shake-flask experiments, the volumetric productivity of 0.4 mg L^{-1} h^{-1} is competitive to algal and yeast-based productions.

2.3.3. Heterologous gene expression for production of C50 carotenoids

On the basis of lycopene-accumulating platform strains, production of a range of C50 carotenoids was established (**Figure 4**). The cyclic C50 carotenoids sarcinaxanthin and C.p. 450 can be derived from lycopene via the heterologous expression of corresponding lycopene elongase and linear C50 carotenoid cyclases. For sarcinaxanthin production genes from the *Micrococcus luteus* [36] were cloned and expressed from a plasmid. The lycopene elongase CrtE2 from this organism adds two DMAPP molecules at C2 and C2′ positions which is identical with the catalytic activity of *C. glutamicum* endogenous lycopene elongase. Simultaneously hydroxylation at the C1/C1′ positions is postulated to stabilise the prenylated carbocation. The linear C50 carotenoid flavuxanthin is cyclised to sarcinaxanthin by a heterodimeric γ-cyclase $CrtY_gY_h$ [36]. This cyclase acts exclusively on a linear C50 backbone and not one lycopene as it holds also true for the ε-cyclase from *C. glutamicum* [36, 49]. Production of C.p. 450 was entailed through plasmid-driven heterologous expression of genes from *Dietzia* sp. CQ4 [65]. Here, elongation of lycopene by LbtBC results in a slightly different linear C50 carotenoid, C.p. 496. Cyclisation by the heterodimeric β-cyclase LbtAB yields C.p. 450. Moreover, starting from C.p. 496, also heterologous synthesis of bacterioruberin could be possible by expression of *cruF*, e.g. from *Haloarcula japonica* [66]; however, this route is not yet established.

2.4. Future prospects: advancing rational strain engineering for short- and long-chain terpenoid production

Based on the findings of the research on carotenoid biosynthesis in *C. glutamicum*, new strategies for rational strain engineering on terpenoid productions are now available. First, deletion of endogenous carotenogenic genes will yield accumulation of the terpenoid precursor molecules DMAPP and IPP in a prophage-cured *C. glutamicum* MB001. Secondly, the application of carbon-chain length-specific prenyltransferases will either yield GPP, FPP or GGPP, allowing their conversion to mono-, di-, sesqui- or polyterpenoids. Besides the C40 and C50 carotenoids, it was also shown that *C. glutamicum* is a suitable host for production of short-chain terpenoids like valencene [28]. Valencene is a sesquiterpene which is derived from FPP. It was shown that heterologous expression of *ispA* from *Escherichia coli* yielded the precursor FPP instead of GGPP as synthesised by native prenyltransferases IdsA and CrtE. Therefore, engineering combinations of native, mutated or heterologous prenyltransferases with terpene synthases from other bacteria or plants are the keys to broaden the terpenoid product spectrum of metabolically engineered *C. glutamicum*.

Since many short-chain terpenoids exhibit antimicrobial properties, timed induction of terpene synthase gene expression is a strategy to face this challenge. An optogenetic approach using photolabile caged IPTG as inducer was successfully applied to allow (i) altered expression levels and (ii) non-invasive timed induction of heterologous genes in a valencene-producing *C. glutamicum* strain [67]. Moreover, recent findings have proven that general transcription machinery engineering (gTME) is an efficient approach to improve carotenoid production in *C. glutamicum* [68]. In this study, carotenoid production was improved in the stationary growth phase either by overexpression of primary sigma factor gene *sigA* or by deletion of alternative sigma factor gene *sigB* [68]. Biosensors have been used in *C. glutamicum* for efficient screening of mutant libraries to find novel targets for metabolic engineering and for positive or negative on-demand control of metabolic pathways or enzymes [64]. An application of a biosensor was already successfully implemented for screening L-lysine-producing *C. glutamicum* strains [69]. Basically, a transcriptional fusion of a regulated promoter and a reporter gene in the presence of the corresponding transcriptional regulator (LysG) enabled intracellular metabolite (lysine) sensing and isolation of lysine-accumulating mutants. On the other hand, intracellular riboswitch-based L-lysine biosensors have been developed to induce the gene of the lysine export system or to repress the citrate synthase gene resulting in increased lysine production [70, 71]. Recently, the CRISPRi/dCas9 system has been established for efficient downregulation of target genes in *C. glutamicum* as exemplified for lysine and glutamate production targets [72]. In summary, a foundation has been laid for employing *C. glutamicum* as a natural terpenoid producer and the recent progress in method and tool development for rational engineering of this biotechnological workhorse foreshadow exciting options for terpenoid production processes using this bacterium.

Author details

Nadja A. Henke, Petra Peters-Wendisch and Volker F. Wendisch*

*Address all correspondence to: volker.wendisch@uni-bielefeld.de

Genetics of Prokaryotes, Faculty of Biology & CeBiTec, Bielefeld University, Germany

References

[1] Gassel S, Schewe H, Schmidt I, Schrader J, Sandmann G. Multiple improvement of astaxanthin biosynthesis in *Xanthophyllomyces dendrorhous* by a combination of conventional mutagenesis and metabolic pathway engineering. Biotechnol Lett. 2013;35(4):565–9.

[2] Lee PC, Schmidt-Dannert C. Metabolic engineering towards biotechnological production of carotenoids in microorganisms. Appl Microbiol Biotechnol. 2002;60(1–2):1–11.

[3] Heider SA, Wendisch VF. Engineering microbial cell factories: metabolic engineering of *Corynebacterium glutamicum* with a focus on non-natural products. Biotechnol J. 2015;10(8):1170–84.

[4] Kinoshita S, Udaka S, Shimono M. Studies on the amino acid fermentation. Production of L-glutamic acid by various microorganisms. J Gen Appl Microbiol. 1957;3:193–205.

[5] Eggeling L, Bott M. A giant market and a powerful metabolism: L-lysine provided by *Corynebacterium glutamicum*. Appl Microbiol Biotechnol. 2015;99(8):3387–94.

[6] Ikeda M. Lysine fermentation: history and genome breeding. Adv Biochem Eng Biotechnol. 2016;1:1–30.

[7] Kalinowski J, Bathe B, Bartels D, Bischoff N, Bott M, Burkovski A, et al. The complete *Corynebacterium glutamicum* ATCC 13032 genome sequence and its impact on the production of L-aspartate-derived amino acids and vitamins. J Biotechnol. 2003;104(1–3):5–25.

[8] Eggeling L, Bott M, editors. Handbook of *Corynebacterium glutamicum*.Boca Raton, USA: CRC Press Taylor & Francis Group; 2005.

[9] Takors R, Bathe B, Rieping M, Hans S, Kelle R, Huthmacher K. Systems biology for industrial strains and fermentation processes-example: amino acids. J Biotechnol. 2007;129(2):181–90.

[10] Kinoshita S, Tanaka K. Glutamic acid. The microbial production of amino acids. Halsted Press: New York, NY, USA; 1972. pp. 263–324.

[11] Blombach B, Seibold GM. Carbohydrate metabolism in *Corynebacterium glutamicum* and applications for the metabolic engineering of L-lysine production strains. Appl Microbiol Biotechnol. 2010;86(5):1313–22.

[12] Meiswinkel TM, Rittmann D, Lindner SN, Wendisch VF. Crude glycerol-based production of amino acids and putrescine by *Corynebacterium glutamicum*. Bioresour Technol. 2013;145:254–8.

[13] Uhde A, Youn JW, Maeda T, Clermont L, Matano C, Kramer R, et al. Glucosamine as carbon source for amino acid-producing *Corynebacterium glutamicum*. Appl Microbiol Biotechnol. 2013;97(4):1679–87.

[14] Matano C, Uhde A, Youn JW, Maeda T, Clermont L, Marin K, et al. Engineering of *Corynebacterium glutamicum* for growth and L-lysine and lycopene production from N-acetyl-glucosamine. Appl Microbiol Biotechnol. 2014;98(12):5633–43.

[15] Tsuchidate T, Tateno T, Okai N, Tanaka T, Ogino C, Kondo A. Glutamate production from beta-glucan using endoglucanase-secreting *Corynebacterium glutamicum*. Appl Microbiol Biotechnol. 2011;90(3):895–901.

[16] Kim EM, Um Y, Bott M, Woo HM. Engineering of *Corynebacterium glutamicum* for growth and succinate production from levoglucosan, a pyrolytic sugar substrate. FEMS Microbiol Lett. 2015;362(19).

[17] Gopinath V, Meiswinkel TM, Wendisch VF, Nampoothiri KM. Amino acid production from rice straw and wheat bran hydrolysates by recombinant pentose-utilizing *Corynebacterium glutamicum*. Appl Microbiol Biotechnol. 2011;92(5):985–96.

[18] Seibold G, Auchter M, Berens S, Kalinowski J, Eikmanns BJ. Utilization of soluble starch by a recombinant *Corynebacterium glutamicum* strain: growth and lysine production. J Biotechnol. 2006;124(2):381–91.

[19] Choi JW, Yim SS, Lee SH, Kang TJ, Park SJ, Jeong KJ. Enhanced production of gamma-aminobutyrate (GABA) in recombinant *Corynebacterium glutamicum* by expressing glutamate decarboxylase active in expanded pH range. Microb Cell Fact. 2015;14(1):21.

[20] Jorge JM, Nguyen AQ, Perez-Garcia F, Kind S, Wendisch VF. Improved fermentative production of gamma-aminobutyric acid via the putrescine route: systems metabolic engineering for production from glucose, amino sugars and xylose. Biotechnol Bioeng. 2016;114(4):862–73..

[21] Eberhardt D, Vold Korgaard Jensen J, Wendisch VF. L-citrulline production by metabolically engineered *Corynebacterium glutamicum* from glucose and alternative carbon sources. AMB Express. 2014;4:85.

[22] Schneider J, Eberhardt D, Wendisch VF. Improving putrescine production by *Corynebacterium glutamicum* by fine-tuning ornithine transcarbamoylase activity using a plasmid addiction system. Appl Microbiol Biotechnol. 2012;95(1):169–78.

[23] Kind S, Neubauer S, Becker J, Yamamoto M, Volkert M, Abendroth G, et al. From zero to hero—production of bio-based nylon from renewable resources using engineered *Corynebacterium glutamicum*. Metab Eng. 2014;25:113–23.

[24] Yamamoto S, Suda M, Niimi S, Inui M, Yukawa H. Strain optimization for efficient isobutanol production using *Corynebacterium glutamicum* under oxygen deprivation. Biotechnol Bioeng. 2013;110(11):2938–48.

[25] Niimi S, Suzuki N, Inui M, Yukawa H. Metabolic engineering of 1,2-propanediol pathways in *Corynebacterium glutamicum*. Appl Microbiol Biotechnol. 2011;90(5):1721–9.

[26] Zahoor A, Otten A, Wendisch VF. Metabolic engineering of *Corynebacterium glutamicum* for glycolate production. J Biotechnol. 2014;192:366–75.

[27] Okino S, Suda M, Fujikura K, Inui M, Yukawa H. Production of D-lactic acid by *Corynebacterium glutamicum* under oxygen deprivation. Appl Microbiol Biotechnol. 2008;78(3):449–54.

[28] Frohwitter J, Heider SA, Peters-Wendisch P, Beekwilder J, Wendisch VF. Production of the sesquiterpene (+)-valencene by metabolically engineered *Corynebacterium glutamicum*. J Biotechnol. 2014;191:205–13..

[29] Henke NA, Heider SA, Peters-Wendisch P, Wendisch VF. Production of the marine carotenoid astaxanthin by metabolically engineered *Corynebacterium glutamicum*. Mar Drugs. 2016;14(7).

[30] Heider SA, Wolf N, Hofemeier A, Peters-Wendisch P, Wendisch VF. Optimization of the IPP precursor supply for the production of lycopene, decaprenoxanthin and astaxanthin by *Corynebacterium glutamicum*. Front Bioeng Biotechnol. 2014;2:28.

[31] Ikeda M, Miyamoto A, Mutoh S, Kitano Y, Tajima M, Shirakura D, et al. Development of biotin-prototrophic and -hyperauxotrophic *Corynebacterium glutamicum* strains. Appl Environ Microbiol. 2013;79(15):4586–94.

[32] Feher T, Papp B, Pal C, Posfai G. Systematic genome reductions: theoretical and experimental approaches. Chem Rev. 2007;107(8):3498–513.

[33] Krubasik P, Takaichi S, Maoka T, Kobayashi M, Masamoto K, Sandmann G. Detailed biosynthetic pathway to decaprenoxanthin diglucoside in *Corynebacterium glutamicum* and identification of novel intermediates. Arch Microbiol. 2001;176(3):217–23.

[34] Kelly M, Jensen SL. Bacterial carotenoids. 26. C50-carotenoids. 2. Bacterioruberin. Acta Chemica Scandinavica. 1967;21(9):2578–80.

[35] Deli J, Matus Z, Molnar P, Toth G, Szalontai G, Steck A, et al. Nigroxanthin (3′,4′-didehydro-beta,gamma-carotene-3,6′-diol), a new carotenoid isolated from paprika (capsicum-annuum var longum-nigrum). Chimia. 1994;48(4):102–4.

[36] Netzer R, Stafsnes MH, Andreassen T, Goksoyr A, Bruheim P, Brautaset T. Biosynthetic pathway for gamma-cyclic sarcinaxanthin in *Micrococcus luteus*: heterologous expression and evidence for diverse and multiple catalytic functions of C(50) carotenoid cyclases. J Bacteriol. 2010;192(21):5688–99.

[37] Saperstein S, Starr MP. The ketonic carotenoid canthaxanthin isolated from a colour mutant of *Corynebacterium michiganense*. Biochem J. 1954;57(2):273–5.

[38] Hodgkiss W, Liston J, Goodwin TW, Jamikorn M. The isolation and description of 2 marine microorganisms with special reference to their pigment production. J Gen Microbiol. 1954;11(3):438–50.

[39] Prebble JN. Carotenoid production in *Corynebacterium fascians* (Tilford) dowson. Biochem J. 1962;85(2):P13–&.

[40] Starr MP, Saperstein S. Thiamine and the carotenoid pigments of *Corynebacterium poinsettia*. Arch Biochem Biophys. 1953;43(1):157–68.

[41] Heider SA, Peters-Wendisch P, Wendisch VF. Carotenoid biosynthesis and overproduction in *Corynebacterium glutamicum*. BMC Microbiol. 2012;12:198(1).

[42] Heider SA, Peters-Wendisch P, Beekwilder J, Wendisch VF. IdsA is the major geranylgeranyl pyrophosphate synthase involved in carotenogenesis in *Corynebacterium glutamicum*. FEBS J. 2014;281(21):4906–20.

[43] Mishra AK, Alderwick LJ, Rittmann D, Wang C, Bhatt A, Jacobs WR, Jr., et al. Identification of a novel alpha(1→6) mannopyranosyltransferase MptB from *Corynebacterium glutamicum* by deletion of a conserved gene, NCgl1505, affords a lipomannan- and lipoarabinomannan-deficient mutant. Mol Microbiol. 2008;68(6):1595–613.

[44] Yang L, Lu S, Belardinelli J, Huc-Claustre E, Jones V, Jackson M, et al. RND transporters protect *Corynebacterium glutamicum* from antibiotics by assembling the outer membrane. Microbiologyopen. 2014;3(4):484–96.

[45] Jungwirth B, Emer D, Brune I, Hansmeier N, Puhler A, Eikmanns BJ, et al. Triple transcriptional control of the resuscitation promoting factor 2 (*rpf2*) gene of *Corynebacterium glutamicum* by the regulators of acetate metabolism RamA and RamB and the cAMP-dependent regulator GlxR. FEMS Microbiol Lett. 2008;281(2):190–7.

[46] Wolf A, Kramer R, Morbach S. Three pathways for trehalose metabolism in *Corynebacterium glutamicum* ATCC13032 and their significance in response to osmotic stress. Mol Microbiol. 2003;49(4):1119–34.

[47] Rossol I, Puhler A. The *Corynebacterium glutamicum aecD* gene encodes a C-S lyase with alpha, beta-elimination activity that degrades aminoethylcysteine. J Bacteriol. 1992;174(9):2968–77.

[48] Schultz C, Niebisch A, Schwaiger A, Viets U, Metzger S, Bramkamp M, et al. Genetic and biochemical analysis of the serine/threonine protein kinases PknA, PknB, PknG and PknL of *Corynebacterium glutamicum*: evidence for non-essentiality and for phosphorylation of OdhI and FtsZ by multiple kinases. Mol Microbiol. 2009;74(3):724–41.

[49] Krubasik P, Kobayashi M, Sandmann G. Expression and functional analysis of a gene cluster involved in the synthesis of decaprenoxanthin reveals the mechanisms for C50 carotenoid formation. Eur J Biochem. 2001;268(13):3702–8.

[50] Heider SA, Peters-Wendisch P, Netzer R, Stafnes M, Brautaset T, Wendisch VF. Production and glucosylation of C50 and C40 carotenoids by metabolically engineered *Corynebacterium glutamicum*. Appl Microbiol Biotechnol. 2014;98(3):1223–35.

[51] Eggeling L, Sahm H. The cell wall barrier of *Corynebacterium glutamicum* and amino acid efflux. J Biosci Bioeng. 2001;92(3):201–13.

[52] Brennan PJ, Lehane DP. The phospholipids of corynebacteria. Lipids. 1971;6(6):401–9.

[53] Britton G. Functions of intact carotenoids. In: Britton, Liaaen-Jensen, Pfander, editors. Carotenoids: natural functions. 4. Basel: Birkhäuser Verlag; 2008. pp. 189–212.

[54] Johnson EA, Schroeder WA. Microbial carotenoids. Adv Biochem Eng Biotechnol. 1996;53:119–78.

[55] Shahmohammadi HR, Asgarani E, Terato H, Saito T, Ohyama Y, Gekko K, et al. Protective roles of bacterioruberin and intracellular KCl in the resistance of *Halobacterium salinarum* against DNA-damaging agents. J Radiat Res. 1998;39(4):251–62.

[56] Abbes M, Baati H, Guermazi S, Messina C, Santulli A, Gharsallah N, et al. Biological properties of carotenoids extracted from *Halobacterium halobium* isolated from a Tunisian solar saltern. BMC Complement Altern Med. 2013;13:255.

[57] Lazrak T, Milon A, Wolff G, Albrecht AM, Miehe M, Ourisson G, et al. Comparison of the effects of inserted C40- and C50-terminally dihydroxylated carotenoids on the mechanical properties of various phospholipid vesicles. Biochim Biophys Acta. 1987;903(1):132–41.

[58] Bansal-Mutalik R, Nikaido H. Quantitative lipid composition of cell envelopes of *Corynebacterium glutamicum* elucidated through reverse micelle extraction. Proc Natl Acad Sci U S A. 2011;108(37):15360–5.

[59] Marchand CH, Salmeron C, Bou Raad R, Meniche X, Chami M, Masi M, et al. Biochemical disclosure of the mycolate outer membrane of *Corynebacterium glutamicum*. J Bacteriol. 2012;194(3):587–97.

[60] Stephens WL, Starr MP. Localization of carotenoid pigment in the cytoplasmic membrane of *Xanthomonas juglandis*. J Bacteriol. 1963;86:1070–4.

[61] Heider SA, Peters-Wendisch P, Wendisch VF, Beekwilder J, Brautaset T. Metabolic engineering for the microbial production of carotenoids and related products with a focus on the rare C50 carotenoids. Appl Microbiol Biotechnol. 2014;98(10):4355–68.

[62] Baumgart M, Unthan S, Ruckert C, Sivalingam J, Grunberger A, Kalinowski J, et al. Construction of a prophage-free variant of *Corynebacterium glutamicum* ATCC 13032 for use as a platform strain for basic research and industrial biotechnology. Appl Environ Microbiol. 2013;79(19):6006–15.

[63] Farasat I, Kushwaha M, Collens J, Easterbrook M, Guido M, Salis HM. Efficient search, mapping, and optimization of multi-protein genetic systems in diverse bacteria. Mol Syst Biol. 2014;10:731.

[64] Henke NA, Heider SAE, Hannibal S, Wendisch VF, Peters-Wendisch P. Isoprenoid pyrophosphate-dependent transcriptional regulation of carotenogenesis in *Corynebacterium glutamicum*. Submitted. 2017.

[65] Tao L, Yao H, Cheng Q. Genes from a *Dietzia* sp. for synthesis of C40 and C50 beta-cyclic carotenoids. Gene. 2007;386(1–2):90–7.

[66] Yang Y, Yatsunami R, Ando A, Miyoko N, Fukui T, Takaichi S, et al. Complete biosynthetic pathway of the C50 carotenoid bacterioruberin from lycopene in the extremely halophilic archaeon *Haloarcula japonica*. J Bacteriol. 2015;197(9):1614–23.

[67] Binder D, Frohwitter J, Mahr R, Bier C, Grunberger A, Loeschcke A, et al. Light-controlled cell factories: employing photocaged isopropyl-beta-D-thiogalactopyranoside for light-mediated optimization of lac promoter-based gene expression and (+)-valencene biosynthesis in *Corynebacterium glutamicum*. Appl Environ Microbiol. 2016;82(20):6141–9.

[68] Taniguchi H, Henke NA, Heider SAE, Wendisch VF. Overexpression of the primary sigma factor gene *sigA* improved carotenoid production by *Corynebacterium glutamicum*: application to production of β-carotene and the non-native linear C50 carotenoid bisanhydrobacterioruberin. Metab Eng Commun. 2017;4:1–11.

[69] Schallmey M, Frunzke J, Eggeling L, Marienhagen J. Looking for the pick of the bunch: high-throughput screening of producing microorganisms with biosensors. Curr Opin Biotechnol. 2014;26:148–54.

[70] Zhou LB, Zeng AP. Exploring lysine riboswitch for metabolic flux control and improvement of L-lysine synthesis in *Corynebacterium glutamicum*. ACS Synth Biol. 2015;4(6):729–34.

[71] Zhou LB, Zeng AP. Engineering a lysine-ON riboswitch for metabolic control of lysine production in *Corynebacterium glutamicum*. ACS Synth Biol. 2015;4(12):1335–40.

[72] Cleto S, Jensen JV, Wendisch VF, Lu TK. *Corynebacterium glutamicum* metabolic engineering with CRISPR interference (CRISPRi). ACS Synth Biol. 2016;5(5):375–85.

Chapter 11

Carotenoids in Yellow Sweet Potatoes, Pumpkins and Yellow Sweet Cassava

Lucia Maria Jaeger de Carvalho,
Gisela Maria Dellamora Ortiz,
José Luiz Viana de Carvalho, Lara Smirdele and
Flavio de Souza Neves Cardoso

Additional information is available at the end of the chapter

http://dx.doi.org/10.5772/67717

Abstract

Carotenoids are the most widespread pigments in nature, extremely important for human health, but are highly unstable molecules especially when exposed to light, oxygen and heat. Many authors report the carotenoid's importance, mainly its pro-vitamin A (α- and β-carotene) and, additionally, the antioxidant capacity of some of them. Currently, more than 600 carotenoids are known and characterized by their chemical structures. In vegetables, common pro-vitamin A carotenoids include β-carotene and its 9, 13 and 15 isomers, α-carotene and β-cryptoxanthin. Other common carotenoids such as lycopene, lutein and zeaxanthin do not have pro-vitamin A activity but serve as natural antioxidants. They are found in many fruits and vegetables such as carrots, yellow sweet potatoes, yellow sweet cassava and pumpkins. Normally, in these plant materials, the β-carotene is the most abundant. It is still used as natural food coloring, which is not very expensive, since enough 3–5 g of β-carotene is used to impart a yellow color characteristic of a ton of margarine. There is also a description of its importance in the formation of compounds responsible for flavors that are of interest fragrance and food industries. The purpose of this chapter is to report the presence of pro-vitamin A carotenoids, mainly the β-carotene in pumpkins, yellow sweet potato and yellow sweet and bitter cassava.

Keywords: carotenoids, cassava, yellow sweet potato, pumpkin, β-carotene

1. Introduction

Carotenoids are the most widespread pigments in nature, extremely important for human health, but are highly unstable molecules especially when exposed to light, oxygen and heat. Many authors report the carotenoid's importance, mainly its pro-vitamin A (α- and β-carotene) and, additionally, the antioxidant capacity of some of them. Currently, more than 600 carotenoids are known and characterized by their chemical structures [1–4]. In vegetables and fruits, common pro-vitamin A carotenoids include β-carotene, α-carotene and β-cryptoxanthin [2, 5]. Other carotenoids such as lycopene are rich in tomatoes. Lutein and zeaxanthin had no pro-vitamin A activity but act as natural antioxidants and are the yellow pigments of the human retinal macula [6–9] and are believed to be responsible for the ophthalmological protective effect of carotenoids, acting as both antioxidants and high-energy blue light filters. In Ref. [10], like spinach, sour cream, corn and egg, lutein and zeaxanthin are the yellow pigments of the human retinal macula [7–9] and are believed to be responsible for the ophthalmological protective effect of carotenoids, acting as both antioxidants and high-energy blue light filters [10, 11]. Carotenoids are found in many fruits and vegetables, such as carrots, yellow sweet potatoes, yellow sweet cassava and pumpkins, among other plant materials. The purpose of this chapter is to make an overview on some aspects of the carotenoids such as sources of pro-vitamin A, functional properties and activities in yellow sweet potatoes, yellow sweet and bitter cassavas and pumpkins.

2. Carotenoids

Nowadays, there is a new approach on the prevention of some pro-vtamin A diseases caused by provitamin A deficiencies, at low costs, mainly the β-carotene. The groups affected by nighblindness, corneal scarring, blindness, measles and, increased mortality in infants (0–12 months), an children (1–6 years) [3] as well as the pregnant and lactating women (blindness) in the underdeveloped and in developed countries or regions around the world.

Years ago, carotenoids like β-carotene were and still is used as natural food colorings, since enough 3–5 g of β-carotene is used to impart a yellow color characteristic in margarines. There is also a description of its importance in the formation of compounds responsible for flavors that are of interest in fragrance and food industries. Besides the food industry, there is the incipient interest of the pharmaceutical industry for its nutritional and functional properties, such as vitamin A precursors and the antioxidant activity [1, 4]. The isomerization is one of the best known chemical properties of carotenoids. Some *cis* (Z) forms can be naturally occurring, as in the case of the first carotenoid biosynthetic, phytoene and phytofluene route, there are 15 predominantly in the *cis (Z)* configuration. Few *cis* forms were found naturally because the *cis* double bond of the presence creates steric hindrance between neighbor groups, making it less stable molecule. Thus, most carotenoids exist in nature in the *trans* configuration, which is more thermodynamically stable [12]. In case of carotenoids, the *cis (Z)* and *trans (E)* designations are determined by the arrangement of substituents of the C=C double bond. Thus, if the substituents are on the same side of the axis, C=C double bond is called *cis*, and if the substituents are on opposite sides of the axis, C=C double bond is called *trans* [12]. The β-carotene naturally occurs in all *trans* (*E*) forms, which is thermodynamically more stable and less soluble as said previously. However, the occurrence of *cis* isomer has been reported, frequently, estimated on the theory, that there are 272 possible isomers of β-carotene, and only 12 were already detected. Among them, easily formed are the *cis* 9, *cis* 13 and *cis*-15. The carotenoids from plant materials contribute with approximately 68% of vitamin A diet globally, and 82% in developed countries. One benefit of pro-vitamin A carotenoids is that they are only converted into vitamin A when the body needs, therefore avoiding its accumulation. On the other hand, several factors influence on its absorption and utilization, such as the type and physical form in the diet, fat intake, vitamin E, fiber contents and the existence of certain diseases and parasitic infections [13]. There are over 600 known carotenoids, some of these compounds are pro-vitamin A and other has little or no vitamin A activity [14].

Carotenoids from vegetables account for 80–85% of dietary vitamin A supply, and their role as a source of pro-vitamin A has attracted great interest due also to the antioxidant potential effect [15].

The biofortification can be an approach to minimize the pro-vitamin A deficiencies and defined as the enrichment of staple crops with essential micronutrients. At present, it is one of the strategies used to alleviate vitamin A deficiency (VAD) by breeding staple crops with β-carotene. Staple crops that have been successfully biofortified with β-carotene under the Harvest*Plus* program are cassava, maize (corn) and sweet potato [16].

Recently, β-apocarotenoids (cleavage products of β-carotene formed by chemical and enzymatic oxidations) were identified and quantified in cantaloupe melons and orange-fleshed honeydew but not in biofortified foods. Biofortified cassava was evaluated; however, there are no detailed analyses of these compounds in biofortified foods and little is known about their bioavailability and intestinal absorption and kinetics of cell uptake and metabolism of β-apocarotenoids. The β-apocarotenoids in roots of non-biofortified cassava varieties were lower than those of biofortified and were hypothesized that these compounds are directly absorbed from the diet similarly to β-carotene (Caco2 cells) [17]. Two unidentified metabolites (X and Y) of β-apo-8′-carotenal. The cellular uptake of β-apo-13-carotenone was rapid, and this compound was extensively degraded over the time. Understanding the mechanisms of absorption and metabolism of β-apocarotenoids relative to their quantities in foods is critical in exploring the functions of these metabolites, some of which have been shown to be potent antagonists of vitamin A.

Hess,Thurnham and Hurrel (2005) [18] reported some studies about the influence of provitamin A carotenoids on status of iron, zinc and vitamin A, considering the effect of β-carotene on vitamin A, requirements by consumption of plant foods, link between vitamin A deficiency and, iron, the possible interactions between vitamin A and, iron metabolism, the link between vitamin A and zinc and, some interventions studies as well the knowledge gap and suggestions for future research. The bioavailability of these micronutrients and their deficiencies in the developed countries in infants, children, and pregnant and lactating women were studied. Another function of some carotenoids like the β-carotene is the protective ability of these pigments that act as antioxidants, acting in preventing peroxidation. Antioxidants are classified into two categories: chain-breaking antioxidants that interfere with the propagation step and preventive antioxidants that interfere with the process initiation step [1]. At high concentrations of O_2, there is a reduction in the antioxidant activity of β-carotene observed in studies conducted in pulmonary tissues. Since peripheral tissues, the efficiency of carotenoids can be greater because the oxygen pressure is lower [19]. The Z isomers of pro-vitamin A have long been known as fitted with vitamin A activity lower than the *E* isomer (*trans*) matching [20]. Furthermore, (*all-E*)-β-carotene was absorbed preferentially to (9-*Z*)-β-carotene in humans [21–23]. The analyses for quantification of total carotenoids are very well described in Ref. [2], using UV/vis spectrophotometry at 450 nm, acetone for extraction and the high-performance liquid chromatography (HPLC) using petroleum ether for their identification. On the other hand, in pro-vitamin A carotenoids, the factors that determine a good antioxidant capacity are as follows: the presence of an electron donor substituent or the hydrogen to radical, depending on their reduction potential; the radical shift capacity formed in its structure and the ability to chelate transition metals involved in oxidative and access to the site of action process, depending on the hydrophilicity or lipophilicity and its partition coefficient [24]. The chemical characteristics of the antioxidants include solubility, regenerative ability, the structure/activity and bioavailability, which are important factors when considering the role of these compounds in human health [25]. In Ref. [26], it was reported that interactions between structurally different compounds and that have variable antioxidant activity promote additional protection against oxidative stress. The antioxidants maybe classified as natural or synthetic. The second antioxidants are widely used in industry, being the most used butylated hydroxyanisole (BHA), butyl hydroxytoluene (BHT), tertiary butyl hydroxyquinone (TBHQ) and propyl gallate (PG). Your choice and concentration vary depending on the food to be used [27–29]. However, due to their potential risks to human health (carcinogenc effects), has increased the interest in research of natural antioxidants [30] present in raw plant materials, processed or not, such as fruit and vegetables, such as tocopherols, ascorbic acid, carotenoids and phenolic compounds [29]. In determination of antioxidant activity of food, in addition to informing its antioxidant potential before ingestion, it is important to assess food's protection against oxidation and deterioration reactions that can lead to decreased quality and its nutritional value [31]. Different methods of measurement of capacity / antioxidant activity of substances and foods such as DPPH (α, α-diphenyl-β-picrylhydrazyl (DPPH) free radical scavenging, ORAC (Oxygen Radical Absorbance Capacity) and ABTS (2,2′-azino-bis(3-ethylbenzothiazoline-6-sulphonic acid). These are shown necessary because of the difficulty of measuring each compound separately and the interactions between the different antioxidants in the system [31–33].

2.1. Sweet and bitter yellow cassava

Cassava (*Manihot esculenta*, Crantz) belongs to the *Euphorbiaceae* family, originated from South America where it was cultivated by the Indians who were responsible for its dissemination in almost all over America. In African, Latin American and Asian continents, it is still one of the main caloric foods to nearly 500 million people, mainly in underdeveloped countries [34, 35]. The variability in total carotenoids and β-carotene and isomers, in 12 varieties of raw yellow bitter cassava roots as well as the degradation in five varieties after the flour process to observe the heat treatment effect on carotenoids degradation, revealed that total carotenoids varied according to the variety, with the β-carotene as the most abundant. At the same time, some varieties presented expressive contents of 13 and 9-*cis*-β-carotene isomers. The total degradation of total carotenoids in the flour, after 19th day of storage, showed the necessity of optimizing the drying process to minimize this loss in order to minimize the deficiency in the Brazilian low-income populations [36]. Studies were conducted in seven raw and cooked roots of yellow sweet cassava to identify total carotenoids, α and β-carotene and its isomers in new varieties that could contribute in the nutritional quality improvement in the populations with malnutrition problems situated in the tropics and, particularly, in the Brazilian Northeast, where the cassava is almost the one of the main cultivations and sometimes the only nutrient source. The *trans*-β-carotene was predominant; however, isomers 13 and 9-*cis* were found in significant quantities compared to the total carotenoids content. However, there was no cooking style that stood out regarding carotenoids retention. Total carotenoids varied in raw roots from 14.15 to 2.64 $\mu g\ g^{-1}$, and total β-carotene from 10.32 to 1.99 $\mu g\ g^{-1}$. The highest content was the all-*E*-β-carotene (4.55 $\mu g\ g^{-1}$). The highest retention % of total carotenoids was found in two varieties (99.49%) and, in total β-carotene (94.31%) were both after cooking. Carotenoids' variability presented the individual potential of the varieties, in the retention prevailed the heat effect in each cooking style applied. However, no cooking style provided a higher retention of total carotenoids or β-carotene uniformly, in all the varieties, with behavior of each variety of sweet yellow cassava roots prevailing in the cooking style. This evaluation showed differences in behaviors that can be attributed to the total carotenoids that were initially found. Differences were found in the cooking styles among the cooking styles regarding total carotenoid and β-carotene in real retention percentage, suggesting and this retention was high for β-carotene [37]. Another aspect that needs to be point out is the storage of cassava roots after harvest for providing important and fundamental information to plant breeding programs aimed at improving cassava storage root nutritional quality. In Ref. [38], Among the 23 cassava landraces with different types of storage, root color and diverse carotenoid types and profiles, the landrace Cas51 (pink color) had low LYCb transcript abundance, whereas landrace Cas64 (intense yellow storage root) had decreased HYb transcript abundance. Lycopene and total β-carotene increased in landraces Cas51 and Cas64, respectively [38]. Thirteen cassava accessions from Brazilian Northeast in two crops were evaluated by many characters. There were accessions identified with potential use as parents in plant breeding to increment of β-carotene BGMC 1221, BGMC 1223 and BGMC 1224 and lycopene BGMC 1222 and BGMC 1223 contents in storage roots [39]. In Ref. [40], the study reported an allelic polymorphism that, in one of the two expressed phytoene synthase (PSY) genes, is capable of enhancing the flux of carbon through carotenogenesis, leading to the accumulation of colored pro-vitamin A carotenoids in storage roots.

2.2. The yellow-flesh sweet potato (*Ipomoea batatas* (L) Lam.)

Ipomoea batatas belongs to the Convolvulaceae family, genus Ipomoea with 50 genera and more than 1000 species. However, it is the most important species and, sometimes, the only staple food crop. The varieties of potatoes with white or pale yellow flesh are less sweet and moist than those with red, pink or orange flesh [41] and are native to the tropical regions in the Americas and Africa [42]. American types with pink yellow flesh contain as high as 5.4–7.2 $mg\ 100\ g^{-1}$ of β-carotene but higher contents. Additionally, more than a dozen African vegetables, this was the richest in folate (1.93–1.96 $mg\ g^{-1}$) [43]. Some cultivars are developed by biofortification program as the yellow/orange-flesh sweet potatoes like the *Beauregard* cultivar in Brazil, with intense orange pulp, because of its high content of β-carotene [16] and, among many studies on it, its antioxidant capacity [44]. The β-carotene is one of the carotenoids with higher pro-vitamin A activity, which is the largest source of vitamin A and its derivatives in the human diet. It is also the most active carotenoid, comprising 15–30% of all serum carotenoids [45]. Because of its high combined structure, the carotenoids are susceptible to degradation by light, oxidation by heat, acid or alkaline pH and the presence of metal ion. They can be hydrophobic, lipophilic, insoluble in water and soluble in solvents such as acetone, alcohol and chloroform. Of more than 600 known carotenoids, only about 50 have pro-vitamin A activity and are antioxidants [46, 47]. On the other hand, studies on the profiles of phenolics, carotenoids and antioxidant capacities of raw and cooked white, yellow, orange, light purple and deep purple sweet potato varieties, grown in Guilin (China), revealed higher anthocyanin contents and antioxidant capacities in purple sweet potato species and higher carotenoid contents in yellow and orange sweet potato. All cooked sweet potatoes exhibited significantly ($p < 0.05$) lower TPC, MAC, TCC, DPPH and Fluorescence recovery after photobleaching (FRAP) values compared to the respective raw samples. Steaming samples showed good results in retention of Total Phenolic Compounds, roasting for keeping anthocyanins, and boiling best preserve the carotenoids [48]. Various types of orange sweet potato (*Ipomoea batatas*) are grown in Brazil and in the world having different shapes and sizes and especially differentiated carotenoid contents of pro-vitamin A. The total carotenoid and β-carotene as well as its isomers 9:13—cis (Z) of β-carotene from two cultivars of orange sweet potato: an organic cultivar called 'carrot', and the *Beauregard* sweet potato Beauregard showed the highest β-carotene content among the studied samples being a good source of provitamin A to be cultivated and consumed, mainly, in the areas of low-income populations and where the deficiency of vitamin A is common among children [16, 44]. *The Beauregard* is a biofortified American cultivar with intense orange pulp because of its high β-carotene content. The effect of the drying treatment on the β-carotene and total carotenoid of this cultivar dried at 40°C for 5 h, 50°C for 2 h and at 60°C for 1 h showed total carotenoids, in mg kg^{-1}, of 129.85 in raw samples; 124.26 in bleached samples; 760.65 (40°C); 769.76 (50°C) and 832.40 (60°C), respectively. The results found by Baganha et al. (2016) for total carotenoids, in mg.kg^{-1}, were 129.85 ± 2.47 in sweet potatoes raw samples; 124.26 ± 3.40 in bleached samples; 760.65 ± 1.45 (40 °C); 769.76 ± 4.43 (50 °C) and 832.40 ± 6.02 (60 °C), respectively. The mean values for β-carotene (mg.kg-1) were 107.93 ± 0.66 (raw); 97.71 ± 4.13 (bleached); 660.08 ± 11.65 (40 °C); 677.03 ± 9.45 (50 °C) and 736.21 ± 3.46 (60 °C), respectively. Drying at 60°C for 1 h showed the highest retention of total carotenoids and β-carotene, indicating that the shortest time of exposure to heat had a greater influence than the higher temperature [44]. In another study, in India, 15 genotypes of exotic and indigenous orange-flesh sweet potatoes cooked were evaluated after cooking process. The β-carotene contents ranged from 28.80 to 97.40 μg g^{-1}, and its retention after cooked varied from 76. 90 to 87.76% [49]. Ten sweet potato clones with different orange flesh color were processed in an oven-drying, boiling, sun-drying and frying. The carotenoids retention depended on the process applied. The highest retentions of total carotenoids and β-carotene were observed in oven-drying (90–91% and 89–96%) followed by boiling (85–90% and 84–90%) and frying (77–85% and 72–86%), and the lowest in both micronutrients were found in the sun-drying method (63–73%) and β-carotene (63–73%) [50, 51]. The extraction step is very important in β-carotene from sweet potato. According to the reference, the best solvent and time of extraction were observed using 91.1% of acetone and 19.6 min of extraction and 278.1 μg g^{-1} of β-carotene in the variety CYY95-26 and small amounts of the isomers 9 and 13-cis [52]. Recently, the β-carotene of four sweet potato varieties from Tanzania (*Jewel, Karoti dar, Kabode* and *Ejumula*) with different intensities of orange flesh color was evaluated. Sweet potatoes were blanched and boiled. There was a threefold reduction in β-carotene content when fresh samples were dried. Boiling results in more retention of β-carotene than blanching in sweet potatoes. The fresh dried had significantly low β-carotene content and low retention on storage compared to boiled and blanched chips, and blanched cowpea leaves retained more β-carotene after 6 months of storage at room temperature [53].

2.3. Pumpkin (*Cucurbita*)

A large number of pumpkin varieties (Cucurbitaceae), each of which containing different amounts of carotenoids, are cultivated worldwide [54]. In Brazil, *C. moschata* cultivars are known to contain high amount of α- and β-carotene. β-carotene has almost 100% pro-vitamin A activity, and α-carotene has approximately 53% pro-vitamin A activity [11, 55–57]. Some varieties such as *C. moschata*, *C. maxima* and *C. pepo*, with color ranging from intense yellow to orange, have revealed high levels of carotenoids, particularly, α and β-carotene, β-criptoxanthin, lutein and zeaxanthin [11]. The orange-fleshed pumpkins (*C. moschata*) normally present high levels of carotenoids mainly β- and α-carotene as well the 9, 13 and 15-β-carotene isomers. Inspite of the low bioaccessibility and bioavailability of the pumpkin carotenoids, its high contents after the cooking styles can still offer adequate daily dietary. On the other hand, the drying process usually can affect the levels of these micronutrients. In Ref. [6], the carotenoid content within pumpkin and squash measured by HPLC and with colorimeter L*a*b* color space values was correlated, and a range of colors and carotenoid types and concentrations within pumpkins and squash was found as well as strong correlations between colorimetric values and carotenoid content were identified. The authors suggested that the genetic variations should make it possible to increase the nutritional value through crossing and selection from within and among the different types with high levels of carotenoids. The α- and β-carotene) of pumpkin flours were evaluated using an oven with air circulation and finally milled in temperatures at 45 and 50°C. Pumpkins are cut into slices, blanched at 90°C for 3 min and dried. The drying process at 45°C spent 132 h (5.5 days) was longer compared with sliced pumpkins dried at 50 (48 hours). In raw pumpkins, total carotenoids were 442.56 μg g^{-1}, α-carotene was 110.87 μg g^{-1}, and β-carotene was 297.37 μg g^{-1}. In flours dried at 45 and 50°C, the total carotenoids were 1892.98 and 1668.43 μg g^{-1}, respectively. Flours presented high contents of carotenoids, as expected, since their moistures were very low (9.17 and 7.83 g 100^{-1}). The flour dried at 45°C preserved 95% of the α-carotene and 83% of the β-carotene compared to the flour dried at 50°C. The isomers 9 and 13-Z- of the β-carotene were present in small percentages in both flours. The results showed to be promising by the fact that the use of these flours in meals in scholar-age children can increase the dietary intake of pro-vitamin A minimizing the vitamin A deficiencies in underdeveloped countries [58]. As wrote previously, some carotenoids are rich in β-carotene, but few are converted by the body into retinol, the active form of vitamin A. These carotenoids are susceptible to degradation (e.g., isomerization and oxidation) during cooking. Total carotenoid, α- and β-carotene, and 9 and 13-Z-β-carotene isomer contents in *C. moschata* after different cooking styles were evaluated. The raw pumpkin presented 236.10, 172.20, 39.95, 3.64 and 0.8610 μg g^{-1} of total carotenoids, β-carotene, α-carotene, 13-*cis*-β-carotene and 9-*Z*-β-carotene, respectively. Samples cooked these total carotenoids in boiling water were 258.50, 184.80, 43.97, 6.80 and 0.77 μg g^{-1}, respectively. Steamed samples revealed 280.77, 202.00, 47.09, 8.23 and 1.247 μg g^{-1}, respectively. Since almost 100% of β-carotene is converted into vitamin A, these results are promising. All carotenoids increased after the cooking methods, most likely of a higher availability induced by the cooking style [59]. The carotenoids should be more bioavailable after the heat treatments. The total carotenoid and β-carotene isomers contents, normally, may increase according to the cooking styles applied. Pumpkin consumption in Northeast Brazil could be more, aggressively, promoted to minimize vitamin A deficiency in this geographic area. Landrace pumpkins occur in nature, and their potential as source of pro-vitamin A, were investigated, in order to be used in conventional plant breeding or biofortification programs, aiming to increase the total carotenoids and β-carotene contents. The total carotenoid, α-carotene, β-carotene and its isomers in two raw landraces pumpkins (*C. moschata*) (A and B) were evaluated to verify its seed production potential. Total carotenoid content of 404.98 (A) and 234.21 μg g^{-1} (B), respectively, were found. The best value for α-carotene contents 72.99 μg g^{-1}. All *E*-β-carotene was the most abundant micronutrient varying from 244.22 to 141.95 μg g^{-1} in both samples. The 9 and 13-Z-carotene isomers were still found in low concentrations. The best β-carotene content in raw sample (A) revealed to be promising for the production of seeds for cultivation and consumption [37]. Recently, the retention of pro-vitamin A carotenoids in the pulp from orange-fleshed pumpkin that was briefly steamed or boiled in either water or water containing 60% sucrose in five genotypes grown in Brazil was investigated and their bioaccessibility in cooked pulp was also determined by *in vitro* digestion and confirmed with Caco-2. Genotypes varied from 209 to 658 μg g^{-1} in pro-vitamin A carotenoids. The retention after cooking was more than 78%. The bioaccessibility of β- and α-carotene was <4%, which showed high variability, affected by food matrix and cooking. One genotype has the potential to provide more than 40% required for children 4–8 years of age per 100 g serving. Pumpkin (*Cucurbita moschata*) is a food crop targeted for enrichment with pro-vitamin A carotenoids [60].

Thus, studies on how the pro-vitamin A carotenoids are assimilated by the human organism, mainly in pumpkins, are relevant and necessary, although, since β-carotene and α-carotene in the pumpkin are poorly bioavailable, these levels are high and supply the daily necessities without the amount of daily food being increased.

Author details

Lucia Maria Jaeger de Carvalho[1*], Gisela Maria Dellamora Ortiz[1], José Luiz Viana de Carvalho[2], Lara Smirdele[1] and Flavio de Souza Neves Cardoso[1]

*Address all correspondence to: luciajaeger@gmail.com

1 Natural Products and Food Department, School of Pharmacy, Federal University of Rio de Janeiro, Rio de Janeiro, RJ, Brazil

2 Embrapa Food Technology, Guaratiba, RJ, Brazil

References

[1] Cardoso S L. Fotofísica de carotenoides e o papel antioxidante de β-caroteno. Química Nova, 20(5): 535–540, 1997. doi:10.1590/S0100-40421997000500014.

[2] Rodriguez-Amaya D B, Kimura M. HarvestPlus Handbook for Carotenoid Analysis. Washington, DC and Cali: IFPRI and CIAT, 58p. HarvestPlus Technical Monograph, 2004.

[3] Stein A J, Meenaskshi Qaim M, Nestel P, Sachdev H P P S, Bhutta Z A. Analyzing the Health Benefits of Biofortified Staple Crops by Means of the Disability-Adjusted Life Years Approach: A Handbook Focusing on Iron, Zinc and Vitamin A. HarvestPlus Breeding Crops for Better Nutrition. Technical Monograph Series, 4. Washington DC and Cali: International Food Policy Research Institute (IFPRI) and International Center for Tropical Agriculture, 2005. 32p.

[4] Uenojo M, Junior M R M, Pastore G M. Carotenoides: propriedades, aplicações e bio transformações para formação de compostos de aroma. Química Nova, 30(3): 616–622, 2007. On-line version ISSN 1678–7064 Printed version ISSN 0100–4042. doi:10.1590/S0100-40422007000300022.

[5] ODS/NIH. Dietary supplement fact sheet: Vitamin A and Carotenoids. http://ods.od/nih.gov/factsheets/vitamina.asp. 2006.

[6] Itle R A, Kabelka E A. Correlation between L*a*b* color space values and carotenoid content in pumpkins and squash (Cucurbita spp.). HortScience, 44(3): 633–637, June, 2009. Print ISSN 0018–5345 and online 2327–9834.

[7] Bone R A, Landrum J T, Fernandez L, Tarzis S L. Analysis of the macular pigment by HPLC: retinal distribution and age study. Investigative Opthalmology & Visual Science, 29: 843–849, 1988.

[8] Handelman G J, Dratz E A, Reay C C, van Kuijk, F J G M. Carotenoids in the human macula and whole retina. Investigative Opthalmology & Visual Science, 29: 850–855, 1988.

[9] Landrum J T and Bone R A. Lutein, zeaxanthin, and the macular pigment. Archives of Biochemistry and Biophysics, 385(1): 28–40, 2001. doi:10.1006/abbi.2000.2171.

[10] Krinsky N I, Landrum J T, Bone R A. Biologic mechanisms of the protective role of lutein and zeaxanthin in the eye. Annual Review of Nutrition, 23: 171–201, 2003. doi:10.1146/annurev.nutr.23.011702.073307.

[11] Rodriguez-Amaya D B, Kimura M, Godoy H T, Amaya-Farfan J. Updated Brazilian database on food carotenoids: factors affecting carotenoid composition. Journal of Food Composition and Analysis, 21: 445–463, 2008. doi:10.1016/j.jfca.2008.04.001.

[12] Penteado M V C. Vitaminas: aspectos nutricionais, bioquímicos, clínicos e analíticos. Barueri, São Paulo: Manole, 2003. doi:10.1590/S1516-93322003000100015.

[13] Souza W A, Boas O M G C V. A deficiência de vitamina A no Brasil: Um panorama. Revista Panamericana de Salud Pública, Washington. 12(3): 173–179, 2002. http://www.scielosp.org/scielo.php?script=sci_arttext&pid=S1020-49892002000900005&lng=en&nrm=iso&tlng=pt, http://iris.paho.org/xmlui/handle/123456789/8524.

[14] Palace V, Hill M H, Farahmand F, Singal P K. Mobilization of antioxidant vitamin pools and hemodynamic function following myocardial infarction. Circulation, 99: 121–126; 1999. doi:10.1161/01.CIR.99.1.12.

[15] Zakaria-Rungkat F, Djaelani M, Setiana, Rumondang E, Nurrochmah. Carotenoid bioavailability of vegetables in carbohydrate-containing food measured by retinol accumulation in rat livers. Journal of Food Composition and Analysis, 13: 297–310, 2000. doi:10.1006/jfca.2000.0871.

[16] EMBRAPA—Empresa Brasileira de Pesquisa Agropecuária. Batata Doce Beauregard: A Batata Vitaminada—Centro Nacional de Pesquisa de Hortaliças. Available in: http://www.cnph.embrapa.br/paginas/produtos/cultivares/batata_doce_beauregard.html.

[17] Durojave B O. β-Apocarotenoids: Occurrence in Cassava Biofortified with β-Carotene and Mechanisms of Uptake in Caco-2 Itestinal Cells. Thesis Presented in Partial Fulfillment of the Requirements for the Degree Master of Science in the Graduate School of The Ohio State University By Graduate Program in Human Nutrition. The Ohio State University. Ohio, USA. 2015.

[18] Hess S Y, Thurnham D I, Hurrel R F. Influence on Pro-Vitamin A Carotenoids on Iron, Zinc, and Vitamin A Status. HarvestPlus Technical Monograph 6. Washington, DC and Calli: International Food Policy Research Institute (IFPRI) and International Center for Tropical Agricultural (CIAT). 2005. 28p.

[19] Cerqueira M F, Medeiros M H G, Augusto O. Antioxidantes dietéticos: Controvérsias e perspectivas. Química Nova, 30(2): 441–449, 2007. doi:10.1590/S0100-40422007000200036.

[20] Zechmeister L. Cis-trans Isomeric Carotenoids, Vitamins A and Arylpolyenes. Vienna: Academic Press, 1962. 1st edition, eBook ISBN: 9781483264936.

[21] Ben-Amotz A, Levy Y. Bioavailability of a natural mixture compared with synthetic all-trans β-carotene in human serum. American Journal of Clinical Nutrition, 63: 729–734, 1996.

[22] Gaziano J M, Johnson E J, Russell R M, Manson J E, Stampfer M J, Ridker P M, Frei B et al. Discrimination in absorption or transport of b-carotene isomers after oral supplementation with either all trans- or 9-cis-b-carotene. American Journal of Clinical Nutrition, 61: 1248–1252, 1995.

[23] Stahl W, Schwarz W, Von Laar J, Sies H. All-trans β-carotene preferentially accumulates in human chylomicrons and very low density lipoproteins compared with 9-cis geometrical isomer. Journal of Nutrition, 125: 2128–2133, 1995. Print ISSN 0022–3166 and online 1541–6100.

[24] Manach C, Scalbert A, Morand C, Rémésy C, Jiménez L. Polyphenols: food sources and bioavailability. The American Journal of Clinical Nutrition, 79(5): 727–747, 2004. Print ISSN 0002–9165 and online 1938–207.

[25] Kaur C, Kapoor H C. Antioxidants in fruits and vegetables—the millennium's health. International Journal of Food Chemistry, 36: 703–725, 2001. doi:10.1111/j.1365-2621.2001.00513.x.

[26] Stahl W, Sies H. Antioxidant activity of carotenoids. Molecular Aspects of Medicine, 24(6): 345–351, 2003. doi:10.1016/S0098-2997(03)00030-X.

[27] Fki I, Allouche N, Sayadi S. The use of polyphenolic extract, purified hydroxytyrosol and 3,4-dihydroxyphenyl acetic acid from olive mill wastewater for the stabilization of refined oils: a potential alternative to synthetic antioxidants. Food Chemistry, 93: 197–204, 2005. doi:10.1016/j.foodchem.2004.09.014.

[28] Han J, Rhee K S. Antioxidant properties of selected oriental non-culinary/nutraceutical herb extracts as evaluated in raw and cooked meat. Meat Science, 70: 25–33, 2005. doi:10.1016/j.meatsci.2004.11.017.

[29] Jardini F A. Avaliação da atividade antioxidante da romã (Punica granatum, L): participação das frações de ácidos fenólicos no processo de inibição da oxidação. Master of Science—Faculdade de Ciências Farmacêuticas, Universidade de São Paulo, Brasil. 2005. 129p.

[30] Cheung LM, Cheung P C K, Ooi V E C. Antioxidant activity and total phenolics of edible mushroom extracts. Food Chemistry, 81: 249–255, 2003. doi:10.1016/S0308-8146(02)00419-3.

[31] Lima A. Caracterização química, avaliação da atividade antioxidante in vitro e in vivo, e identificação dos compostos fenólicos presentes no pequi (Caryocar brasiliense Cambi). PhD. Thesis—Faculdade de Ciências Farmacêuticas, Universidade de São Paulo, Rio de Janeiro, Brazil. 2008. 219p.

[32] Kulkarni A P, Aradhya S M, Divakar S. Isolation and identification of a radical scavenging antioxidant punicalagin from pith and carpellary membrane of pomegranate fruit. Food Chemistry, 87: 551–557, 2004. doi:10.1016/j.foodchem.2004.01.006.

[33] Scherer R, Godoy H T. Antioxidant activity index (AAI) by the 2, 2-diphenyl-1-picrylhydrazyl method. Food Chemistry, 112(3): 654–658, 2009. doi:10.1016/j.foodchem.2008.06.026.

[34] Maduagwu E N, Okorowkwo C O, Okafor P N. Occupational and dietary exposures of humans to cyanide poisoning from large-scale cassava processing and ingestion of cassava foods. Food and Chemical Toxicology, 40: 1001–1005, 2002. doi:10.1016/S0278-6915(01)00109-0.

[35] Oliveira A R G. Avaliação e estudo da retenção de carotenoides totais e β-caroteno em mandioca amarela mansa e brava. Evaluation and study of total carotenoids and β-carotene in yellow sweet and bitter cassava. Master Science. Rio de Janeiro, Brazil: Rural University of Rio de Janeiro, 2005.

[36] Oliveira A R G, Carvalho L M J, Nutti M R, Carvalho J L V, Fukuda W G. Assessment and degradation study of total carotenoid and β-carotene in bitter yellow cassava varieties. African Journal of Food Science, 4(4): 148–155, April 2010. Available from: http://www.academicjournals.org/ajfs. ISSN 1996–0794 © 2010 Academic Journals.

[37] Carvalho L M J, Gomes P B, Godoy R L O, Pacheco S, Monte P H F, Carvalho J LV, Nutti M R, Neves A C L, Vieira A C R A, Ramos S R R. Total carotenoid content, α-carotene and β-carotene, of landrace pumpkins (Cucurbita moschata Duch): a preliminary study. Food Research International, 47(20): 337–340, July, 2012. doi:10.1016/j.foodres.2011.07.040.

[38] Carvalho L J, Agustini M A, Anderson J V, Vieira E A, de Souza C R, Chen S, Schaal B A, Silva J P. Natural variation in expression of genes associated with carotenoid biosynthesis and accumulation in cassava (Manihot esculenta Crantz) storage root. BMC Plant Biology, 16(1): 133, June, 2016. doi:10.1186/s12870-016-0826-0.

[39] Silva K N, Vieira E. A, Fialho J F, Carvalho L J C B, Silva M S. Agronomic potential and carotenoid contents within cassava storage roots. Ciência Rural, 44: 8, 2014. doi:10.1590/0103-8478cr20130606.

[40] Welsch R, Arango J, Bär C, Salazar B., Al-Babili S, Chavarriaga J B P, Ceballos H, Tohme J, Beyer P. Pro-vitamin A accumulation in cassava (Manihot esculenta) roots driven by a single nucleotide polymorphism in a phytoene synthase gene. 2010 American Society of Plant Biologists. The Plant Cell, 22(10): 3348–3356, October 2010. doi:10.1105/tpc.110.077560.

[41] Loebenstein G, Thottappilly B V G. The sweet potato. Springer Business Media, 2009, pp 391–425. ISBN 978-1-4020-9475-0. doi:10.1007/978-1-4020-9475-0_8.

[42] Duke, 1983

[43] Hup R S, Abalaka JA, Stafford W L. Folate content of various Nigerian foods. Journal of Food and Agriculture, 34(4): 404–406, 1993. doi:10.1002/jsfa.2740340413.

[44] Baganha C L, Fernandez A A, Santos E P, Karse I M, Chern M S, Santos Y C S, Carvalho J L V, Mello A F S, Carvalho L M J, Cabral L M.2015. Efeito do processamento térmico nos teores de β-caroteno e de carotenoides totais em batata-doce de polpa alaranjada biofortificada. In: V Jornada Integrada de Pós-graduação da Área de Farmácia da UFRJ. 29–30 September, 2016.

[45] Gomes F S. Carotenoides: uma possível proteção contra o desenvolvimento de câncer. Revista de Nutrição, 20(5), Oct. 2007. doi:10.1590/S1415-52732007000500009.

[46] Ambrosio C L B, Campos F A C S, Faro Z P. Carotenoides como alternativa contra a hipovitaminose A. Revista de Nutrição, 19(2): 233–243, April, 2006. doi:10.1590/S1415-52732006000200010 On-lineversion. ISSN:1678–9865.

[47] Zaccari F, Giovanni G, Soto B, Las R. Color y contenido de β-carotenos en boniatos, crudos y cocidos, durante su almacenamiento en Uruguay. Agrociencia Uruguay, Montevideo, 16(1), June 2012. On-line ISSN 2301–154.

[48] Tang Y, Cai W, Xu B. Profiles of phenolics, carotenoids and antioxidative capacities of thermal processed white, yellow, orange and purple sweet potatoes grown in Guilin. Food Science and Human Wellness, 4: 123–132, 2015. doi:10.1016/j.fshw.2015.07.003.

[49] Mitra S. Nutritional status of orange-fleshed sweet potatoes in alleviating vitamin A malnutrition through a food-based approach. Journal of Nutrition and Food Science, 2: 160, 2012. doi:10.4172/2155-9600.1000160.

[50] Vimala B, Sreekanth A, Binu, H, Gruneberg W. Variability in 42 orange-fleshed sweet potato hybrids for tuber yield and carotene and dry matter content. Gene Conserve (Brazil), 10(41): 190–200, 2011. ISSN 1808–1878.

[51] Vimala B, Nambisan B A, Hariprakash B. Retention of carotenoids in orange-fleshed sweet potato during processing. Journal of Food Science and Technology, 48: 520–524, 2011. doi:10.1007/s13197-011-0323-2.Publishedonline2011Apr3.

[52] Lien Ching-Yi, Chan Chin-Feng, Huang Che-Lun, Lai Yung-Chang, Liao W C. Studies of carotene extraction from sweet potato variety CYY95-26, Ipomoea batatas, L. International Journal of Food Engineering. 8(2): 1556–3758, 2012. ISSN (Online). doi: 10.1515/1556-3758.2490.

[53] Nicanuru C. Effect of pretreatments and drying on nutrient content of orange fleshed sweet potato tubers and cowpea leaves used in Maswa District, Tanzania. Master of Science (Food Science and Technology). Jomo Kenyatta University of Agriculture And Technology, Gramado, Rio Grande do Sul, Brazil. 2016. 69p.

[54] Mínguez-Mosquera M I, Hornero-Méndez D, Pérez-Gálvez A. Carotenoids and pro-vitamin A in functional foods. In W J Hurst (Ed.), Methods of analysis for functional foods and nutraceuticals. Washington: CRC Press, 1, 2001. pp 101–158. doi:10.1201/9781420014679.ch3.

[55] Silva S R, Mercadante A Z. Composição de carotenoides de maracujá-amarelo (Passiflora edulis flavicarpa) in natura. Ciência e Tecnologia de Alimentos, 22: 254–258, 2002. doi:10.1590/S0101-20612002000300001.

[56] Boiteux L S, Nascimento W M, Fonseca M E N, Lana M M, Reis A, Mendonça J L, Lopes J F, Reifschneider F J B 'Brasileirinha': cultivar de abóbora (Cucurbita moschata) de frutos bicolores com valor ornamental e aptidão para consumo verde. Horticultura Brasileira, 25(1): 103–106, 2007.

[57] Carvalho L M J, Minguita A P S, Carvalho J L V, Ramos S R, Barbosa D C M. Estimation of β and α -carotene in oven dried pumpkins slices for biofortified flours. 2016. In: Book of Abstracts. 29th EFFoST International Conference. Food Science Research and Innovation. Delivering sustainable solutions to the global economy and society (EFoST 2015). From 10–12 November, 2015. Atenas, Greece.

[58] Carvalho L M J, Smiderle L A S M, Carvalho J L V, Cardoso F S N, Koblitz M G B. Assessment of carotenoids in pumpkins after different home cooking conditions. Food Science and Technology, 34(2). Campinas April/June 2014. On-line version ISSN 1678-457Xdoi:10.1590/fst.2014.0058.

[59] Ribeiro E M G, Chitchumroonchokchai C, Carvalho L M J, Moura F F, Carvalho J L V, Failla M L. Effect of style of home cooking on retention and bioaccessibility of pro-vitamin A carotenoids in biofortified pumpkin (Cucurbita moschata Duch.). Food Research International, 77(3): 620–626, November 2015. doi:10.1016/j.foodres.2015.08.038.

[60] Carvalho L M J, Carvalho J L V, Faustino R M E B, Kaser I M, Lima V G, Sousa D S F. Variability of total carotenoids in C. moschata Genotypes. Chemical Engineering Transactions, 44: 247–252, 2015 doi:10.3303/CET1544042.

Additional References Consulted:

[1] Carvalho L M J, Baganha C L, Carvalho J L V, Chern M C, Santos Y S, Paiva E, Mello A F S, Barbosa D M C, Gomes P B. Diferenciação de carotenoides totais em cultivares comuns, orgânicas e, linhagens de batata doce de polpa laranja. In: XXV Congressso Brasileiro de Ciência e Tecnologia de Alimentos (CBCTA)—Alimentação: a árvore que sustenta a vida and X CGIAR Section IV International Symposium—Food the tree that sustains life. Gramado, Brasil. From 24–27 October 2016. ID. 1669. http://www.ufrgs.br/sbctars-eventos/xxvcbcta/anais/ Accessed in November, 2016. http://www.ufrgs.br/sbctars-eventos/xxvcbcta/anais/. ISBN online 978-85-89123-06-8.

[2] Purseglove J W. Tropical crops: dicotyledons. New York: Wiley, 1968. 719 p (pp. 80, 81 and 88. Editor Longman).

[3] Purseglove J W. Tropical crops: dicotyledons. Science, 163(3871): 1050–1051, 1969. doi:10.1126/science.163.3871.1050.

[4] Lucia C M D, Campos F M, Mata G M S C, Sant'ana H M P. Controle de perdas de carotenoides em hortaliças preparadas em unidades de alimentação e nutrição hospitalar. Ciência & Saúde Coletiva, 13(5): 1627–1636, 2008. doi:10.1590/S1413-81232008000500026.

[5] Pacheco S. Preparo de padrões analíticos, estudo de estabilidade e parâmetros de validação para ensaio de carotenoides por cromatografia líquida. Seropédica: UFRRJ, 2009. 106p. Master in Food Science and Technology.

[6] Scott GJ, Best R, Rosegrant M, Bokanga M. Roots and Tubers in the Global Food System: A Vision Statement to the Year 2020. International Potato Center, and others, 2000. ISBN 92-9060-203-1.

Chapter 12

Carotenoids in Cassava Roots

Hernán Ceballos, Fabrice Davrieux, Elise F. Talsma,

John Belalcazar, Paul Chavarriaga and

Meike S. Andersson

Additional information is available at the end of the chapter

http://dx.doi.org/10.5772/intechopen.68279

Abstract

Vitamin A deficiency (VAD) is a preventable tragedy that affects millions of people, particularly in sub-Saharan Africa. A large proportion of these people rely on diets based on cassava as a source of calories. During the last two decades, significant efforts have been made to identify sources of germplasm with high pro-vitamin A carotenoids (pVAC) and then use them to develop cultivars with a nutritional goal of 15 μg g^{-1} of β-carotene (fresh weight basis) and good agronomic performance. The protocols for sampling roots and quantifying carotenoids have been improved. Recently, NIR predictions began to be used. Retention of carotenoids after different root processing methods has been measured. Bioavailability studies suggest high conversion rates. Genetic modification has also been achieved with mixed results. Carotenogenesis genes have been characterized and their activity in roots measured.

Keywords: carotenogenesis, conventional breeding, dry matter content, physiological deterioration

1. Introduction: cassava as an important food-security and industrial crop

Cassava (*Manihot esculenta* Crantz) has a Neotropical origin and significant economic relevance. It is an important crop in tropical and subtropical regions of the world, growing from sea level up to 1800 m. Its most common product is the starchy root, but cassava foliage has an excellent nutritional quality for animal and human consumption and offers great potential [1, 2]. Stems are used for commercial multiplication. Therefore, every part of the plant can be used and exploited. Cassava is the fourth most important basic food after rice, wheat, and maize worldwide, but is the second most important food staple (in terms of calories consumed) in sub-Saharan Africa [3–6]. The crop is called Africa's food insurance because it offers reliable yields even in the face of drought, low soil fertility, low intensity management, and because of its resilience to face the effects of climate change [7–9].

Between 2010 and 2014, an average of 22 million ha was annually grown with cassava worldwide (70% in Africa, 18% in Asia, and about 12% in the Americas). The area planted to cassava has grown steadily, and it was 2011 when, for the first time, more than 20 million ha were planted with the crop (FAOSTAT). Markets in cassava are diverse. The crop was initially domesticated for the direct consumption of the roots, which contain plenty of carbohydrates. However, roots contain little protein and few micronutrients [10], when compared to sweet potatoes, potatoes, beans, maize, or wheat. Globally, in the period 1970–2003, the main uses of cassava roots were for food (54%), followed by feed (30%), and other uses including starch production (4%) [11]. Global use of cassava for feed was affected by the reduction of imports from the European Union in the 1980s. Production of starch, on the other hand, increased considerably in the same period (by 17.5% annually [11]). Today, cassava is the second most important source of starch worldwide [12]. In the 2000s, a considerable amount of cassava roots started to be used for the production of fuel ethanol as well [13, 14].

A comprehensive screening of root quality traits from more than 4000 cassava clones has been published [15]. On average, roots from these genotypes had 33.6% of dry matter content (DMC) of which 84.5% is starch (the main and most valuable product in cassava roots). Cassava starch quality is excellent and has, on average, 20.7% amylose (the remaining 79.3% is amylopectin). Cassava roots spoil quickly (2–3 days after harvest) because of a process called postharvest physiological deterioration (PPD). Therefore, roots need to be processed or consumed soon after harvest.

Vitamin A is an essential micronutrient for the normal functioning of the visual and immune systems, growth and development, maintenance of epithelial cellular integrity, and for reproduction [16]. Vitamin A deficiency (VAD) is the leading cause of preventable blindness in children and increases the risk of disease and death from severe infections. In pregnant women, VAD causes night blindness and may increase the risk of maternal mortality. VAD is a public health problem in more than half of all countries, especially in Africa and South-East Asia, hitting hardest young children (visual impairment and blindness, and significantly increases the risk of severe illness, and even death) and pregnant women (especially during the last trimester when demand by both the unborn child and the mother is highest) in low-income countries. An estimated 250 million preschool children show VAD, and it is likely that, in areas where VAD is prevalent, a substantial proportion of pregnant women is also affected. An estimated 250,000 to 500,000 vitamin A-deficient children become blind every year, half of them dying within 12 months of losing their sight [17].

Vitamin A exists in natural products in many different forms: as preformed retinoids (stored in animal tissues) and as provitamin A carotenoids (pVAC), which are synthesized as pigments by many plants and are found in different plant tissues [16]. The carotenoids are present in both plant and animal food products (in animal products their occurrence results from dietary exposure). Retinoids, on the other hand, are only found in animal products. A comprehensive review on carotenoids has been recently published [18].

Different strategies (dietary diversification, food fortification, and/or supplementation) and considerable efforts have been made to reduce VAD worldwide [19]. These strategies are relatively cost-effective, but have failed to completely eradicate the problem for a diversity of reasons [20]. Moreover, prevalence of VAD has remained unchanged in sub-Saharan Africa and south Asia during the 1991–2013 periods [21]. Recently, different programs such as HarvestPlus (www.harvestplus.org), involving a global alliance of research institutions, initiated the implementation of a fourth strategy (biofortification) to develop micronutrient-dense staple crops [19, 22, 23]. A diet rich in pVAC, in addition to reducing the problems related to VAD would also result in other health benefits, including a reduction of cancer incidence [24, 25].

The color of root parenchyma (e.g., pulp) in cassava varies from white to yellow. Pinkish pigmentation has also been reported [26]. Pigmentation of the parenchyma is closely linked to carotenoids content. Most cassava varieties worldwide produce roots with white parenchyma. This is particularly appreciated by the starch industry. However, breeding to develop cassava varieties need to target different end users and it is only recently that the efforts to develop high-carotenoids biofortified germplasm were initiated by HarvestPlus. This is an international, interdisciplinary research initiative that seeks to reduce human malnutrition by increasing micronutrients in staple crops, including cassava. The progress already attained by HarvesPlus and the awareness that has elicited in the plant breeding community resulted in the 2016 World Food Prize Award. Biofortified cassava makes sense considering the importance of this crop in sub-Saharan Africa and the reported prevalence of VAD in that region of the world. The target for these varieties will be not only for human consumption but also for animal feed, particularly poultry [27].

2. Nutritional value of cassava roots

The nutritional quality of cassava roots in general is low, and contains mainly carbohydrates. Per 100 gram raw weight, white cassava provides 160 kcal mainly as carbohydrates (38 g) and contains further water (60 g), a little protein (1.4 g), fat (0.3 g) and trace elements of iron (0.3 mg), niacin 0.9 mg), thiamin (0.09 mg), riboflavin (0.05 mg), calcium (16 mg), potassium (271 mg), zinc (0.3 mg), and vitamin C (21 g) [28]. People depending on a diet predominantly based on white cassava roots are at greater risk of having iron, zinc and VAD, as was shown in children in Kenya and Nigeria [29].

Cassava also contains variety-dependent concentrations of cyanogenic glucosides (CG), which are toxic for humans and needs to be eliminated during processing before consumption. Therefore, cassava with high concentration in cyanide are better suitable for more processed products like porridge made out of flour and those low in cyanide are more suitable for boiled consumption [30]. CG levels range from low (≤100 ppm) in roots from sweet/cool varieties to very high (≥3000 ppm) in bitter cassava cultivars. CG are eliminated through alternative processing techniques [2]. However, health of people may be affected [31], particularly in years when drought has affected other crops, which increases people's dependency in cassava. In addition to higher consumption of cassava roots, drought generally increases CG levels in them [32].

3. Genetic variation of carotenoids content in cassava roots

Carotenoids perform essential physiological functions in plants. They are involved in the photosynthesis process, protecting the plant from photo-oxidative damage, and as precursors of regulatory molecules, such as abscisic acid (ABA) and strigolactones [33, 34]. In cassava, carotenoids can be found in leaves and roots. Concentration is much larger in leaves than in roots [35–37]. The earliest reports on pVAC contents in cassava roots (written in Portuguese) were based on cassava samples from the Amazon region of Brazil and the earliest dates back to 1964 [38–40]. Interest in increasing pVAC content in cassava roots began in the 1990s in India [41, 42] and few years later at International Center for Tropical Agriculture (CIAT) in Colombia [43] and at IIITA in Nigeria [37]. HarvestPlus provided financial resources and encouraged systematic work that resulted in the screening of a large sample of cassava germplasm [26, 36].

Quality of data improved over the years. Initially most of the information focused on assessing differences in the intensity of parenchyma pigmentation [44] and often relied on measuring total carotenoids content (TCC) by spectrophotometer. HPLC analysis restricted the number of samples that could be screened. Moorthy reported in 1990 a range of variation for β-carotene from negligible to 7.9 $\mu g\ g^{-1}$ [41]. Unless otherwise specified, all concentration values for carotenoids will be expressed on fresh weight basis. A comprehensive screening of carotenoids content reported in 2005 (1789 genotypes) had an average total carotenoids content (TCC) of 2.46 $\mu g\ g^{-1}$, ranging from 1.02 to 10.40 $\mu g\ g^{-1}$ [26]. **Figure 1** illustrates the strong skewness in frequency distributions for TTC by 2005 with a long tail to the right (low frequency of genotypes with high TCC values).

The basic concept in breeding is to cross outstanding genotypes (e.g., clones) to generate a new generation of segregating progenies that will hopefully have a better average performance. At CIAT those individuals, at the right of the plot in **Figure 1**, were crossed among themselves to obtain botanical seed. Each seed represented a new, genetically unique, genotype. The seed was germinated and the roots from the plants generated were evaluated for their carotenoids content. The best materials (highest pVAC) were then selected and crossed to produce a new cycle of selection. The basic scheme was described by Ceballos and coworkers in 2013 [45]. Very early in this process, it became clear that carotenoids content was not only closely associated with pigmentation of the root parenchyma [41, 43] but also with a reduction or delay of PPD [26, 46, 47]. The beneficial effect of increased TCC in lengthening the shelf life of cassava roots was an important finding that would encourage the adoption

http://dx.doi.org/10.5772/intechopen.68279

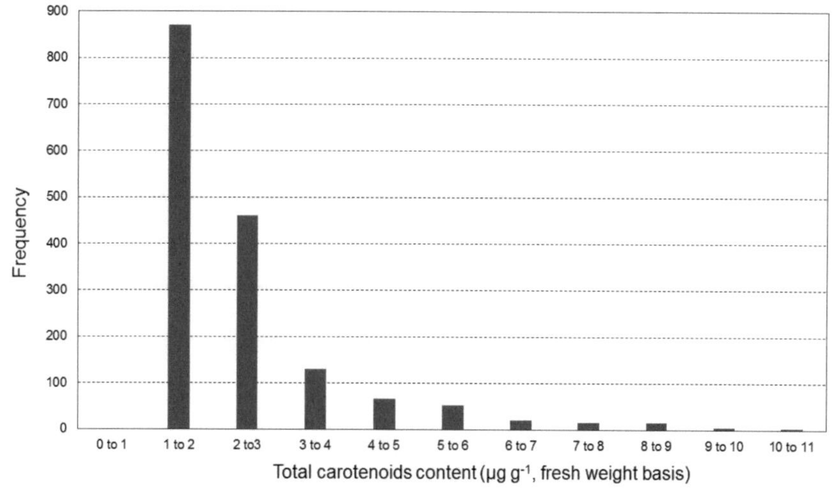

Figure 1. Frequency distribution for total carotenoids content (μg g^{-1}, fresh weight basis) of cassava germplasm by 2005.

of biofortified varieties. It has been suggested that reduced PPD in high pVAC roots may involve β-ionone-like molecules, derived from β-carotene catabolism, which play a role in the response to biotic stress such as fungal infection [48]. However, PPD is a very variable trait, difficult to measure visually, influenced by the environment, and which depends very much on the storage conditions of the harvested root.

The relationships depicted in **Figure 2** are relevant for the impact of the biofortified varieties released through the HarvestPlus initiative. It is not enough to develop cassava cultivars with high pVAC. The roots from these cultivars should meet consumer preferences. Key parameters defining consumer acceptance are dry matter content (DMC) and cyanogenic potential (particularly in regions of the world where roots are boiled). DMC influences texture after boiling and is also a key parameter in the production of gari, for example (a popular way to consume cassava in Africa). The relationship between carotenoids and DMC is basically nonexistent (**Figure 2**). It is possible, therefore, to identify genotypes with high pVAC and acceptable levels of DMC. The first target of the HarvestPlus initiative was to develop biofortified clones for Africa. The key related trait for the most important ways to consume cassava in Africa would be DMC. Cyanogenic potential is not critical for gari production, but it is critical for table consumption after boiling. This is the most common way to consume cassava in many Latin American and Caribbean (LAC) countries. Only recently CIAT started the development of biofortified cassava with low cyanogenic potential targeting LAC. The relationship between carotenoids and cyanogenic potential in **Figure 2** shows a negative trend, although the coefficient of determination is low ($r^2 = 0.15$). For a variety to be considered "sweet" and apt for table consumption, the maximum HCN levels would be about 150 ppm. At this stage of the breeding process, only a few dozen genotypes have been found to have low HCN, high pVAC and acceptable cooking quality. Efforts are currently made in making crosses among these genotypes to increase the number of segregating materials that can fit the consumer preferences in LAC.

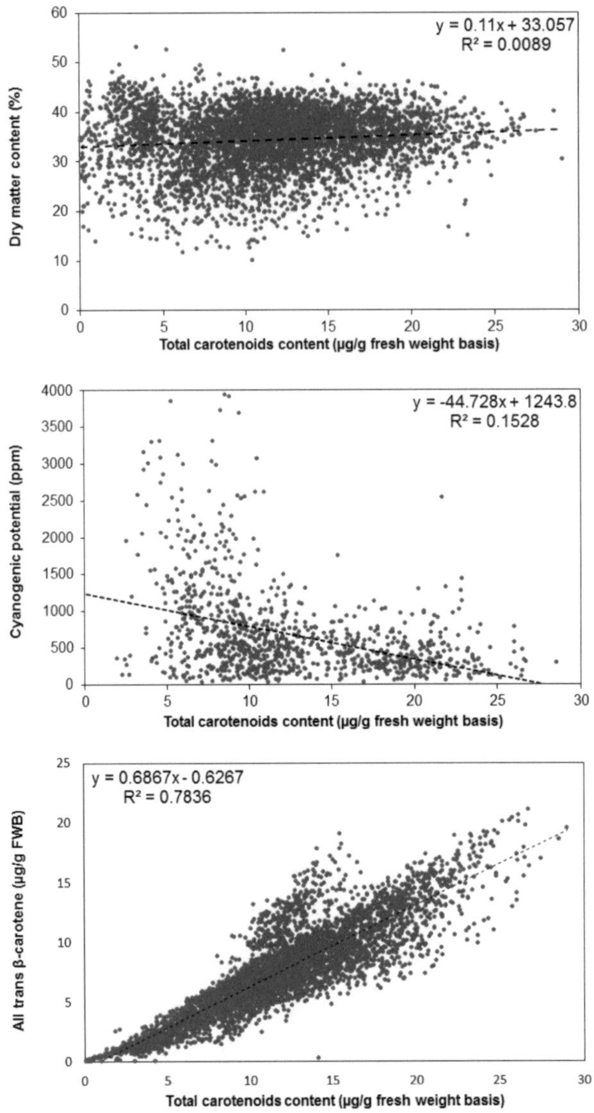

Figure 2. Relationship between total carotenoids content (μg/g, fresh weight basis) with dry matter content (TOP); cyanogenic potential (MIDDLE) and all-trans β-carotene (bottom) contents.

About 600 carotenoids have been isolated and characterized in nature, and approximately 10% of these can be metabolized into vitamin A by mammals. The most important carotenoids with vitaminic activity are β- and α-carotenes and cryptoxanthins. Some carotenoids that cannot be converted into vitamin A (e.g., lutein, zeaxanthin, and lycopene) can be found in the parenchyma of cassava root as well. Not all pVAC carotenoids have the same activity. β-carotene has about twice as much vitamin activity as the remaining pVAC carotenoids [49]. Fortunately, as illustrated in **Figure 2**, most of carotenoids in cassava roots are β-carotene. Assessing the nutritional potential of cassava roots, therefore, requires partitioning TCC into its different individual carotenoid components. This is usually done through HPLC chromatograms.

Table 1 presents information on key root quality traits such as DMC and HCN and a description of the different carotenoids quantified in roots from a large sample of cassava genotypes. The information presented in **Table 1** is clearly unbalanced with large variation in the number of samples used for quantifying different parameters. For example, DMC is available for 4913 samples, whereas HCN was only measured in 981genotypes. Data for all-trans β-carotene is available from 4952 chromatograms, whereas only 49 samples allowed measuring α-carotene. To a large extent, the limited information on α-carotene is due to the low concentration, often below detection level, observed for this carotenoid. It is clear from data presented in **Table 1** that most carotenoids present in cassava roots are β-carotene (TBC), with a prevalence of its all-trans isomer. Similar conclusions were reported by Maziya-Dixon and Dixon in 2015 [50].

	Count	**Average**	**(Standard deviation)**	**Range**		**Skew**
Dry matter content (%)	4913	34	(±6)	10	–53	–1
HCN total (ppm)	981	684	(±631)	22	–3927	2
Precursors in α- and β-carotene synthesis μg g^{-1}, fresh weight basis						
Phytoene	3992	4.12	(±3.10)	0.00	–22.3	1.17
Phytofluene	2635	2.12	(±1.87)	0.00	–12.2	1.36
Carotenoids concentrations measured in μg g^{-1}, fresh weight basis						
Total carotenoids (spectroph.)	4922	11.59	(±5.16)	0.07	–29.5	0.14
Total carotenoids (HPLC)	4952	11.51	(±5.08)	0.11	–29.0	0.12
β-carotene and derived carotenoids μg g^{-1}, fresh weight basis						
All-trans β-carotene	4952	7.28	(±3.94)	0.00	–21.0	0.31
15-cis β-carotene	4107	0.15	(±0.15)	0.00	–2.99	2.18
13-cis β-carotene	4920	1.01	(±0.60)	0.00	–2.24	0.74
9-cis β-carotene	4895	0.99	(±0.54)	0.00	–3.95	0.82
β -Criptoxanthin	4074	0.14	(±0.13)	0.00	–3.77	1.20
Anteroxanthins	4038	0.30	(±0.26)	0.00	–1.46	2.03
Violaxanthins	4469	0.38	(±0.25)	0.00	–3.74	1.12
α-carotene and derived carotenoids μg g^{-1}, fresh weight basis						
α-carotene	49	0.08	(±0.01)	0.04	–1.09	–0.22
Lutein	4853	0.38	(±0.41)	0.00	–0.10	2.17

Table 1. Root quality traits and carotenoid components quantified in a large sample of cassava roots.

4. Inheritance of carotenoids content in cassava roots

Early attempts to explain the inheritance of carotenoids content in cassava roots were actually based on the visual assessment of intensity of pigmentation in root parenchyma, which was not always linked to the quantification of carotenoids content [44, 51]. Iglesias and co-workers [43] suggested in 1997 that relatively few major genes were involved, based on a hypothesis of two genes with epistatic effects controlling root color. One of these genes would show complete dominance while the second would have partial dominance. These early studies typically describe three classes for the intensity of root pigmentation (white, cream and yellow parenchyma) and agree that the trait shows dominance and is controlled by few genes, thus suggesting high heritability. Akinwale and co-workers [52] also suggested that segregations in TCC can be explained by the genetic control of two genes showing complete dominance.

During the past few years, several studies have been published on the heritability of carotenoids content in cassava roots, but based on their systematic quantification, rather than indirectly assessing the intensity of pigmentation. Parent-offspring regression analysis confirmed high heritability (>0.60) for the trait [53–55]. High heritability for carotenoids content in cassava is further confirmed in stability studies assessing the relative importance of genotype, environment, and genotype-by-environment interaction [56–58] and estimations of general and specific combining ability effects [59]. An analysis of the segregation for TCC in large full-sib and self-pollinated families was conducted aiming at identifying patterns that could be explained by Mendelian genetics [53, 54]. **Table 2** reproduces the segregating values in 14 self-pollinated (S_1) families. It is clear that there is an association between the TCC values of the progenitors and the average of their respective S_1 progenies (further validating the high heritability of the trait). The range of variation in the resulting S_1 progenies is wide, and in most cases, there are individuals with TCC values above those of the respective progenitors (except for one family that only had two individuals). Based on the results of these segregations, it was postulated again that there are at least two genes explaining the high, intermediate, and low TCC values [53, 54].

However, one of these genes (*I*-) would drastically reduce the accumulation of carotenoids and shows complete dominance (one copy of the gene is enough to reduce TCC levels). Breeders may be interested in the homozygous recessive genotype for this gene (*ii*). The other gene (*C*-), would contribute to carotenoids accumulation and shows partial dominance.

The large variation for TCC values in the S_1 family AM702 is worth highlighting (**Table 2**). This family had 29 individual genotypes and the range of variation for TCC was from 0.63 to 19.1 μg/g. The performance in family AM697 is also interesting because none of its 38 members had TCC values above 1.0 μg/g. The progenitors of these two families are related to each other since they were siblings from the same full-sib family (GM708). It has been proposed that the progenitor of AM697 was homozygous for the allele reducing TCC accumulation (*II*) and/or was lacking the allele that promoted the accumulation of carotenoids (*cc*) [53, 54].

Family	TCC in progenitor	n	Minimum	Maximum	Average	Standard deviation	Skewness
S_1 family from genotypes derived from full-sib family CM9816							
AM690	9.99 (H)	90	2.89	13.10	6.51	1.82	−0.599
AM689	6.67 (I)	71	0.46	11.37	4.73	2.69	−0.045
AM692	5.04 (I)	48	4.85	8.95	6.38	0.77	0.026
AM691	1.87 (L)	73	0.40	5.89	2.34	1.18	0.892
S_1 family from genotypes derived from full-sib family GM708							
AM702	11.16 (H)	29	0.63	19.1	6.46	5.92	0.734
AM700	7.32 (I)	2	2.01	2.79	2.40	0.55	n.d.
AM698	0.54 (L)	34	0.30	6.56	1.10	1.25	3.298
AM697	0.27(L)	38	0.26	0.89	0.54	0.15	0.34
S_1 family from genotypes derived from full-sib family GM893							
AM710	8.77 (H)	29	4.85	8.95	6.38	0.77	−0.299
AM718	8.14 (H)	38	3.19	9.24	5.90	1.28	−0.025
AM712	6.06 (I)	57	0.44	8.41	3.31	2.06	0.478
AM720	2.04 (L)	40	0.14	9.72	2.28	2.22	0.432
S_1 families derived from commercial clones MTAI 8 and CM 4919-1							
AM320	2.14 (L)	177	0.37	8.10	2.22	1.37	1.084
AM324	3.83 (I)	11	0.75	8.58	3.83	2.35	0.313

Table 2. Descriptive statistics in roots from S_1 families derived from progenitors selected because of their high (H), intermediate (I) or low (L) levels of TCC (μg/g fresh weight basis).

It may be convenient to relate the three intensities of pigmentation used in earlier studies to the TCC values of the progenitors in **Table 2**. A root with white parenchyma has TCC values up to 1.5–2.0 μg/g, those with a cream pulp may show TCC values ranging from 1.5 to 3.0 μg/g, whereas a TCC value above 3.0–3.5 μg/g is observed only in yellow roots. It is useful to visualize two levels of variation in carotenoids content in cassava roots. The first level of variation is **qualitative** and relates to the three phenotypes regarding intensity of pigmentation (e.g. white, cream, or yellow roots with the TCC values mentioned above). It is expected that this qualitative variation relates to some of the genes in the carotenoids biosynthesis described below. There is, however, a second level of variation that is **quantitative** in nature. It involves genotypes only with yellow roots and covers a very wide range of variation (from 3.0 to 30.0 μg/g). Understanding the genetic factors controlling the large quantitative variation for TCC among yellow-rooted cassava is critical for breeding cassava with higher levels of pVAC. The distinction between white, cream and yellow roots, on the other hand, is easy to make visually and has already been linked to some of the genes in the carotenogesis pathway.

The pathway in carotenoids biosynthesis has been known for more than three decades. However, it was only after the 1990s that the genes involved in it could be cloned [60]. The diagram in **Figure 3** illustrates the key steps in the process and is helpful for understanding the inheritance of carotenoids synthesis and their accumulation, as crystals, in root chromoplasts. Few conclusions can be drawn from **Figure 3**: (a) The carotenogenesis pathway is relatively simple with few key steps (and therefore few genes); (b) breeding should downregulate the activity of lycopene ε-cyclase (LCYE) to favor the synthesis of β-carotene on the right side of **Figure 3**; and (c) breeding should try to reduce catabolic conversion of β-carotene into zeaxanthin, which eventually leads to the production of abscisic acid (ABA).

Phytoene synthase (PSY) is the enzyme responsible for the synthesis of phytoene, which is the first reaction specifically related to the carotenoid synthesis pathway as illustrated in **Figure 3** and demonstrated in cassava [34], also see later in this chapter. There are three copies of PSY in cassava but transcripts for one of them (PSY3) were negligible [61]. It has been proposed that PSY1 is mostly involved in responses to stress (through ABA), whereas PSY2 would be involved in carotenoids synthesis and accumulation [61]. Genetic transformation works to enhance carotenoids content in cassava roots through the simultaneous over-expression of PSY and CRTI (a bacterial version of phytoene desaturase—PDS) have been successful. Transformed cassava produced yellow roots with considerably higher TCC values compared

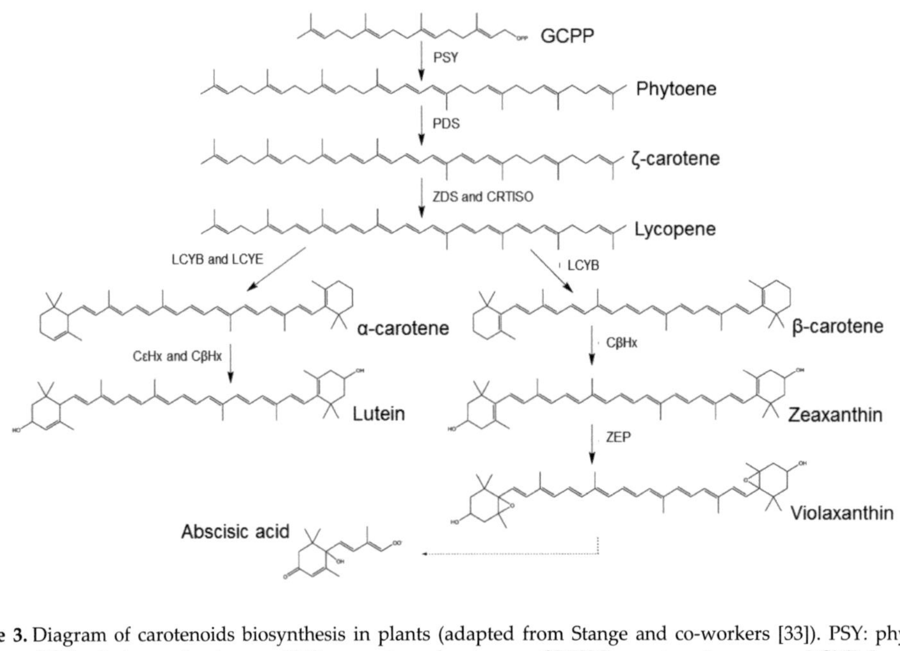

Figure 3. Diagram of carotenoids biosynthesis in plants (adapted from Stange and co-workers [33]). PSY: phytoene synthase; PDS: phytoene desaturase; ZDS: z-carotene desaturase; CRTISO: carotene isomerase; LCYE: Lycopene ε-cyclase; LCYB: Lycopene β cyclase; CεHx: ε-carotene hydroxylase; CβHx: β-carotene hydroxylase; ZEP: zeaxanthin epoxidase.

with the untransformed version of the same genotype that produces roots with white parenchyma [34, 62, 63]. It is clear, therefore, that PSY explains the variation between white and yellow roots (e.g., TCC values below 2.0 μg/g versus those > 3.0 μg/g). It is reasonable to hypothesize that the partially dominant factor (C-) in Morillo-Coronado and co-workers [54] is related to PSY.

There is no need to deregulate the activity of LCYE in cassava because the pathway naturally favors the accumulation of β-carotene as demonstrated by the values presented in **Table 1**. In fact, only trace values of α-carotene and lutein are found in cassava roots. Downregulation of β-carotene hydroxylases (*CβHx*), the third conclusion above, on the other hand, may be desirable. Reduced activity for this enzyme has been shown to enhance β-carotene concentrations in potato and sweet potato [64, 65]. No information has been published regarding variation for *CβHx* in cassava. However, the putative recessive trait in phenotypic studies (*ii*) may be related to a reduced activity of this enzyme [54].

Molecular markers ranging from microsatellite (SSR) to single nucleotide polymorphisms (SNP), identifying regions in the genome responsible for carotenoids content have been reported as well [34, 66–70]. It would be expected that some of the QTLs identified in these studies are related to the PSY gene. In fact, that was the case reported by Esuma and co-workers in 2016 [67]. Other markers may be linked to other genes in the carotenogenesis pathway (perhaps the putative *I* factor mentioned in Morillo-Coronado et al. article [54]. which may be related to *CβHx*). These genes are likely linked to the qualitative variation mentioned above (e.g., distinguishing white, cream, and yellow roots). From the nutritional point of view, however, much more relevant is the quantitative variation observed among genotypes producing yellow roots (ranging from 3.0 to 30.0 μg/g). Interesting studies in carrot compared roots from three cultivars: one producing white roots with negligible amounts of carotenoids crystals (CC) and two cultivars producing yellow roots with vast differences in CC [71]. In this study, the authors concluded that the difference in CC among the two cultivars producing yellow roots was not due to increased numbers of carotene-containing chromoplasts but rather greater accumulation of carotene per chromoplast. In other words, the carotenogenesis pathways were similar in both cultivars producing yellow roots, but the chromoplasts of the cultivar bred for higher levels of CC show a higher demand or capacity to store them. Perhaps future molecular work in cassava should focus in identifying QTLs related to the *sink strength* of chromoplast in their demand for carotenoids and/or capacity to store them as crystals. These studies should only focus on cassava genotypes producing yellow roots ranging, for example, from 10 to 30 μg/g of TCC.

The relationship between carotenoids and ABA (**Figure 3**) is relevant. ABA has been shown to have important effects as a plant growth regulator [61, 72]. As illustrated in **Figure 2**, there is no relationship between TCC and DMC in the germplasm screened at CIAT. However, many studies in Africa report a clearly negative correlation between the two traits [61, 72]. Perhaps African breeding populations have restricted genetic variability (particularly for *CβHx*) and some of the β-carotene initially produced is further converted into ABA. This hypothesis would be supported by a genetic transformation work in which, parallel to an increase in TCC, there is a reduction in DMC and increase in ABA. The performance of six transformed genotypes with high pVAC has been reported [63]. TCC in the wild type was 0.38 μg/g, whereas the best transgenic line showed 5.73 μg/g. Similarly, TBC values increased from 0.12 μg/g in the wild type to 4.67 μg/g in the best transgenic line. One distinctive feature of the transgenic genotypes is a drastic reduction of DMC. The wild-type roots had 32.6% DMC, whereas the average DMC of the six transgenic lines was 20.2% (ranging from 15.8 to 23.8%). There seems to be a generalized fact that increases of pVAC through genetic transformation result in a (pleiotropic) decrease in DMC. A reasonable explanation for the simultaneous increase in pVAC and reduction of DMC in the transgenic lines would be that the pathway did not stop at the step accumulating β-carotene but continued resulting in a higher production of ABA.

5. Evolution and improvements of sampling protocols for measuring carotenoids in cassava roots

Only a small tissue section (5 g) is needed for extracting and quantifying carotenoids in cassava roots. At the inception of the HarvestPlus initiative, there was no information regarding the uniformity of pVAC concentration within the root, among roots of the same plant, among roots from different plants of the same genotype, and on the relative importance of environment, age, and genotype-by-environment interaction. For early work, roots were cut longitudinally in four sections: two diagonally opposed sections were used for DMC determination and the remaining two quarters were chopped into small pieces which were thoroughly mixed, and from this bulked tissue, a random sample was used for pVAC quantification.

The first systematic study analyzing sampling variation of pVAC concentration in a cassava clone that produces yellow roots (average TCC 3.90 μg/g) was published in 2008 [73]. Samples along the longitudinal (proximal, central, and distal section) and across the transversal (periphery, mid-parenchyma, and core) axes of the roots were analyzed. Carotenoids and dry matter content were quantified individually in 243 root samples. Average TCC values, on a fresh weight basis, were higher in the proximal sections (4.10 μg/g) and gradually lower in the central (3.86 μg/g) and distal portions (3.73 μg/g). An opposite trend was observed when carotenoids were quantified in dry weight basis. Carotenoids concentrations were higher in the core (4.13 μg/g) and lower toward the mid-parenchyma (4.04 μg/g) and the periphery of the root (3.52 μg/g), both fresh and dry weight basis. Plant-to-plant variation was only significant for dry matter content.

Breeding progress was successful increasing pVAC levels. Along this process, however, laboratory personnel began to observe an increase in the variability of intensity of pigmentation within the root and among roots from the same genotype, even when coming from the same plant. **Figure 4** illustrates some cases with clear differences in intensity of pigmentation particularly across the root. It should be pointed that these differences are not always present and the bottom left photograph shows a case with a more uniform pigmentation. A new and more complete study on sampling variation was published in 2011 [74]. In this new study of many genotypes, most of them with considerably higher levels of pVAC compared with the earlier

Figure 4. Illustration of longitudinal (left) and transversal (center) variation in the intensity of parenchyma pigmentation. On right: (**A**): chopped roots; (**B**): homogenized paste after processing roots with a food processor; (**C**): illustration of variation in root color in a full-sib family.

work [73] were analyzed. Variation in aliquot quantifications from the same root was negligible indicating a reliable experimental procedure. A large source of variation for carotenoids was due to differences among the 26 genotypes analyzed (ranging from 2.87 to 12.95 μg/g). In contrast to earlier results, root-to-root variation from the same plant was surprisingly high in some cases and accounted for an average of 25% of the total variation. Plant-to-plant variation was not as high and accounted for 20% of the total variance. Carotenoid content was shown to vary depending on the age of the plant as well, particularly comparing samples harvested at 8 and 10 months after planting. Single-plant evaluations for carotenoid content in cassava, which is a requirement for rapid cycling recurrent selection (described later in this chapter) was acceptable, considering that it reduces in half the time required for evaluation and selection. However, it was suggested that two to three roots per plant are combined together in a sample to better represent each genotype at a standard plant age (10–12 months after planting).

CIAT laboratory began using an industrial-grade food processor that quickly grinds the root samples in to a uniform paste (**Figure 4B**). This approach allowed overcoming the problem of variation in pVAC concentration along and across the root (which resulted in variation in the coarsely chopped pieces, **Figure 4A**), as well as the operational problems of having to use two to three roots per genotype. Early studies made apparent that the variation in DMC along the root influenced the concentrations of pVAC [73]. Variation of DMC across the root, on the other hand, seemed to be less relevant. The influence of variation in DMC in different sections of the root on the quantification of carotenoids was also shown in Ref. [75]. It is important to recognize, therefore, that there is not always a linear relationship between pVAC concentrations reported on a fresh and a dry weight basis. Since DMC is generally around 30–35%, the rule of the thumb dictates that the relationship between TTC and TBC reported on a dry weight basis should be around threefold higher than when reported on a fresh weight basis. However, DMC can vary considerably (from 10 to 50% as shown in **Table 1**) in experimental material. **Figure 5** has been presented to illustrate how a pVAC concentration of 15 μg/g (on a fresh weight basis) will vary when expressed on a dry weight basis, depending on the DMC values of the root. It is important for researchers reporting data on pVAC concentrations to also provide the respective levels of DMC.

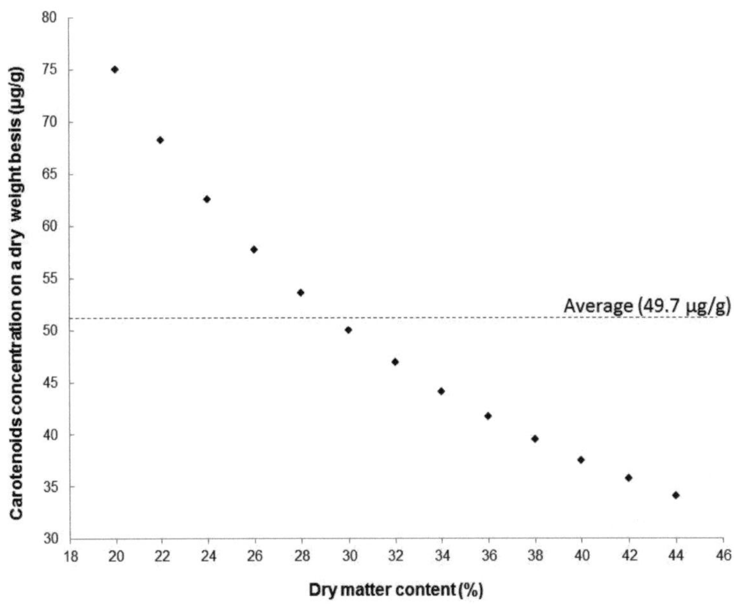

Figure 5. Illustration of how DMC in the root will affect the concentration of carotenoids expressed on a dry weight basis in a sample that showed 15 μg/g of pVAC on a fresh weigh basis. When DMC is low (22%) pVAC on a dry weight basis tends to be very high (75 μg/g), whereas roots with higher DMC (a desirable trait) show lower pVAC values, if expressed on a dry weight basis (e.g., <45 μg/g).

6. Evolution and improvements of quantifying protocols for carotenoids in cassava roots

Early work quantifying carotenoids content relied on standard spectrophotometry and HPLC quantifications. Adjustments in the extraction protocol these techniques require, however, had to be made [76, 77].

One of the adjustments was in the separation of the solid and liquid phases that was carried out by centrifugation and not by filtration [26]. Carotenoids are sensitive to ultraviolet (UV) light, pro-oxidants or associated compounds, and high temperature. Thus, steps need to be taken to avoid any adverse changes in this pigment due to such effects protecting them from UV light and avoiding excessively high temperatures. Therefore, special care to avoid direct exposure of ground tissue samples to sunlight is critical. Likewise, the lights in the laboratory need to be protected with UV filters. The typical extraction protocol requires 5 g of root tissue (either fresh or boiled) which is added to a vial with 10 mL acetone. After 10 minutes, 10 ml of petroleum ether are added and mixed using an ultra-turrax for 1 min. Samples are then centrifuged at 3000 RPM, for 10 min, at 10 °C. The organic phase is collected and extraction repeated on the residue with 5 ml of acetone and 5 ml of petroleum ether, followed by centrifugation. Extractions are optimized until it residues turn colorless. Based on preliminary analysis, it was decided that three iterative extractions would be used for fresh root samples and four for boiled samples. The organic phases are combined with 10 mL of 0.1 M NaCl solution and centrifuged (3000 RPM, for 7 min, at 10 °C). This washing process is repeated two additional times. The aqueous phase is extracted with a pipette. Petroleum ether is added to the extracts to adjust volume to 15 ml.

With the extracts obtained, TCC can be determined by visible absorption spectrophotometry, at an absorbance at 450 nm and using the absorption coefficient of β-carotene in petroleum ether (2592) [76, 77]. Partitioning of TCC into concentrations of individual carotenoids is done by HPLC. From the organic phase used for spectrophotometric quantification of TCC, aliquots (15 mL) are taken and partially dried by nitrogen evaporator. Immediately before injection, the dry extract is dissolved in 1 mL of (1:1) methanol and methy tert-butyl ether HPLC-grade and filtered through a 0.22 μm PTFE filter. Separation and quantification of carotenoids are achieved at CIAT using an YMC Carotenoid S-5 C30 reversed-phase column (4.6 mm × 150 mm: particle size, 5 μm), with a YMC carotenoid S-5 guard column (4.0 × 23 mm) in a HPLC, using DAD detector with wavelength set at 450 nm. Peaks are identified by comparing retention time and spectral characteristics against a pure standard and available literature [78].

Breeding projects to increase carotenoids content require screening hundreds of samples in a short period of time. Quantification of carotenoids by spectrophotometry or HPLC limits the number of samples that can be analyzed. Therefore, a large data set linking TCC and TBC data with Near-infrared spectroscopy (NIRS) spectra was gradually developed.

NIRS is based on the absorption of electromagnetic radiation at wavelengths in the range of 780–2500 nm. The interaction between the electromagnetic radiations and the vibrational properties of the chemical bonds results in absorption of a part of the radiation energy. The electromagnetic spectrum is divided into several regions, each of which induces specific molecular or atomic transition and is therefore suited to a specific type of spectroscopy. NIR spectroscopy belongs to the class of methods called vibrational spectroscopy techniques. This class of techniques aims to analyze a product in order to obtain qualitative and/or quantitative information. The principle of absorption can be interpreted as a resonance phenomenon: when the vibrational frequency of a specific chemical bond is equal to the frequency of the infrared radiation, a part of the energy is absorbed by the chemical bond. NIR spectra of foods comprise broad bands arising from overlapping absorptions corresponding mainly to overtones and combinations of vibrational modes involving C-H, O-H, and N-H chemical bonds [79]. Additional information on the theory regarding vibrational NIR spectroscopy is found in more detail in several Refs [80].

The key point is that a NIR spectrum is the resultant of all the elementary absorptions due to the chemical constituents of the product analyzed; the spectrum is as a fingerprint of the sample. This fingerprint contains qualitative and quantitative information about the physical and chemical composition of the sample. Due to this complexity, spectra are treated mathematically in order to extract the relevant information within the spectra linked to the property of interest (carotenoids or other). This step, called calibration, aims to develop mathematical models, which link the reference values to a linear combination of the values of absorbance. The calibration step is based on chemometric methods that applies multivariate analyses such as multiple linear regression (MLR), partial least squares (PLS), or principle components analysis (PCA) to the spectra in order to quantify the property analyzed. Thus, NIRS is a secondary, indirect method, and the calibration step requires a primary method that provides the value of the property for each samples. Once the calibration is developed and validated, it can be applied to new samples and used to directly quantify (actually predict) the property from the spectrum.

In the routine analyses, NIRS method is simple, nondestructive, and rapid, with minimum sample preparation and environmentally friendly. The NIR technique is widely used in the agriculture sector. However, a few studies have been conducted on nutritional properties of fresh tubers or roots using NIR spectroscopy. McGoverin and collaborators inventoried studies on carotenoids in various crops and one related to carotenoids in potato [81]. The efficiency of NIR spectroscopy for predicting TCC, TBC, and DMC in fresh cassava roots has been demonstrated [82]. The models were based on partial least squares (PLS) regression. PLS regression ranged within the linear methods, which assume that the relationship between the independent and dependent variables are linear in nature. However, predictions of a new harvest based on the PLS models were actually "extrapolations" because, year after year, cassava genotypes with higher carotenoids content were obtained by the breeding project. This resulted in a nonlinear response and a tendency to underestimate the highest contents. The use of LOCAL regression algorithm based on large database has circumvented this restriction [83].

The cassava database (6026 samples) has been built over 6 years (2009–2014). For each genotype, two to three commercial-size roots were taken to the lab where they were washed, peeled, and homogenized with a food processor into a homogenous paste. Further analyses were made using aliquots from this homogeneous paste. For NIR analysis, approximately 8 g of ground root tissue was placed in NIR spectroscopy capsules for analysis using a FOSS 6500 monochromator with autocup sampling module. All spectra were recorded from 400 to 2498 nm at 2 nm intervals and saved as the average of 32 scans. Each sample was duplicated. Therefore, spectra from two root subsamples were obtained per genotype. Each of these two samples was measured once. Further analyses were made on the average of the two spectra available per genotype. Spectra were corrected for light scattering using the standard normal variate and de-trend (SNVD) correction. Then, the second derivative of the Log(1/R) spectrum, calculated on five data points and smoothed using Savitzky-Golay polynomial smoothing on five data points, was used in combination with LOCAL regressions to develop prediction models [84].

The laboratory analyses led to 4277 TCC values that ranged from 0.11 to 29.0 μg g^{-1} with an average value of 11.6 μg g^{-1} and 4288 TBC values that ranged from negligible to 20.1 μg g^{-1} with an average value of 6.9 μg g^{-1}. The standard deviation (SD) was 5.1 μg g^{-1} for TCC and 3.6 μg g^{-1} for TBC. All values were measured and expressed on a fresh weight basis. The DMC values (n = 5578) ranged between 12.3 and 52.4% with a SD of 5.9%. Between 2009 and 2014, increases in TCC and TBC were 86 and 122%, respectively. The standard error of prediction were 1.38 μg g^{-1} for TCC and 1.02 μg g^{-1} for TBC and 1.09% for DMC using LOCAL regression. The scatter plots of NIRS values versus HPLC values for TBC and TCC illustrate the high performances of the models (**Figure 6**). The multiple determination coefficients were higher than 0.9 for both constituents.

After 5 years of harvest and database building, NIR spectroscopy coupled with LOCAL regression led to accurate and robust calibrations for breeding programs aiming at increasing carotenoids content in fresh cassava roots [85]. These results offer immense prospects for many cassava-breeding projects in the world; NIRS overtakes the bottleneck of conventional carotenoids quantification methods. Classically, in a well-equipped laboratory with experienced personnel, 20–30 samples per day can be analyzed; the implementation of NIRS in the analytical chain boosts this number by five to ten. Moreover, the possibility to share data and calibrations between spectrometers make it possible to develop a network for high-throughput phenotyping of fresh cassava roots.

An alternative quantification protocol has been implemented at the International Institute of Tropical Agriculture (IITA) in Nigeria and other cassava research programs in Africa. This method is based on the iCheck™ Carotene technology [86].

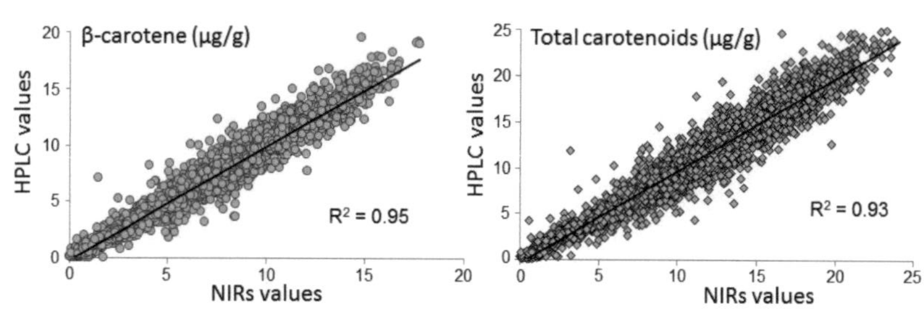

Figure 6. Scatter plots of TBC (left plot) and TCC (right plot) NIR spectroscopy values versus HPLC values.

7. Evolution and improvements of breeding methods to increase carotenoids in cassava roots

Cassava breeding relies on a method known as phenotypic recurrent selection. Although it is a simple approach, each cycle of selection requires about 8 years for completion. Elite clones are crossed to produce full- or half-sib families. In the former, the identity of both (male and female) progenitors is known. In the latter, only the female progenitor is known [78]. Large number of botanical seed from the crosses of elite germplasm is generated each year. The seeds are then germinated to produce seedling plants. At this stage, called F1, only one plant per genotype is available. The seedling plants are grown for 11–12 months (the standard age for harvesting commercial cassava) when they are selected based on different traits such as vigor, plant architecture, resistance/tolerance to pest and diseases, and/or starch and root quality traits. Stems of selected plants are harvested and used as a source of planting material for the next stage of selection (single-row trials—SRT). Typically, eight stem cuttings are used to represent each genotype in SRT, which is the first stage where cloned plants are evaluated. The rate of vegetative multiplication (number of cuttings that can be obtained from a given plant) is relatively low in cassava (1:8 to 1:10) and, therefore, it takes several years to have enough planting material from a given genotype to be evaluated in multi-location trials.

Following SRT are the preliminary yield trials (PYT), advanced yield trials (AYT), and uniform yield trials (UYT), which are usually conducted for two consecutive years. Each of these types of trials increases the size of the plots, the number of replications, and/or the number of locations used in the evaluation process. Usually, 4000–5000 seeds are germinated, and about 4000 seedlings transplanted for the F1 stage. The number of genotypes in SRT, PYT, ADYT, and UYT is gradually reduced from about 2500, 200, 60, and 20, respectively. The entire process takes about 8 years for completion. This lengthy process is necessary because yield data is prone to large experimental errors and affected by genotype-by-environment interaction.

Breeding for increased levels of pVAC, however, is much easier because of the high heritability of the trait. Breeders do not need data from replicated trials using large plots in many locations to identify a genotype with higher levels of pVAC. In fact, a single plant is enough as results from studies on sampling variation demonstrated. Therefore, a special breeding approach—rapid cycling recurrent selection RCRS—was implemented [45]. In this scheme, seedling plants were evaluated for carotenoids content and the best genotypes preselected based on single plant evaluations. Preselection was based on a visual assessment for the intensity of pigmentation that discarded genotypes producing roots with white, cream, or pale yellow roots. Yellow roots from genotypes preselected in the field were then sent to the laboratory for carotenoids quantification. Selection was based primarily on TCC/TBC levels but other traits such as DMC and root yield potential were also taken into account. Selected genotypes were immediately incorporated into the crossing blocks to be used as progenitors.

Each recurrent selection cycle, therefore, lasted 2–3 years (depending on how quickly the selected materials flowered). Selected genotypes were also incorporated into the normal selection process described above (SRT, PYT, AYT, and UYT) to identify genotypes that not only had excellent levels of pVAC but also acceptable to outstanding agronomic performance. Number of genotypes involved was different compared with ordinary cassava breeding. The F1 seedling stage was considerably larger with 8000 to 10,000 seeds germinated and about 5000 to 8000 plants grown through the season. About 1000 to 1500 plants were selected in the field and about 500–800 of them eventually screened for pVAC levels in the laboratory. This breeding approach was very successful and resulted in three to fourfold increases in TCC and TBC in the maximum values observed at the F1 trials in a decade [45]. This kind of genetic progress was in fact unprecedented in cassava and is largely due to the high heritability of the trait.

RCRS, however, faced some limiting bottleneck in its initial scheme. Selection for TCC/TBC of seedling plants required extraction and quantification of pVAC that was time-consuming. Only six to eight samples per day could be analyzed through HPLC. This resulted in many logistic problems that were gradually identified. The harvesting season lasted up to 4 months rather and 2–3 weeks. This was necessary to screen at least 500–800 samples in the laboratory. Extending the harvesting season for such a long period implied that some genotypes were harvested during the dry season and others after the arrival of the rains. This, in turn had some impact on DMC of the roots because this parameter is highest at the end of the dry season but is considerably reduced after the rains began. The availability of irrigation at CIAT reduced the difference in dry and wet season, but only partially. As the problem of a lengthy harvesting season became evident, CIAT began the development of the protocol for predicting pVAC and DMC based on NIRS, as described above [82, 84]. The possibility of selecting for high pVAC based on reliable NIRS predictions was a major breakthrough. It allowed increasing the number of samples analyzed (from 500–800 to 2000–2500), while reducing the harvesting season (from 4 months down to 3–4 weeks).

An improved RCRS scheme has been recently described [85]. In this scheme, seedling plants (F1 stage) are grown only for 6 months. Since plants are young, only three vegetative cuttings can be taken from selected genotypes. These cuttings are planted to grow a new stage (F1C1) that was not used previously. In the F1C1, each genotype has been cloned and is represented by three plants. Selection is conducted in two steps. The visual assessment done in the field in the old system is still done at the F1 stage, with the difference that it is done when harvested plants are only 6-months old. Only genotypes producing yellow roots are selected, but other traits such as resistance to thrips, adequate vigor, and acceptable yield potential are also taken into consideration. A key feature is that seedlings are transplanted off-season, and the harvest of the plants takes place during the normal harvesting/planting season. Therefore, planting of the F1C1 is done in the usual season. The F1C1 is then grown for a full season and plants harvested at 10–12 months of age. However, in the new system, three plants per genotype are available.

One of these plants is harvested at the end of the dry season for NIRS quantification of pVAC and DMC. Stems from selected genotypes can be used for planting a new crossing block. The remaining two plants of selected genotypes are left in the field and are used as a source of planting material for further phenotypic selection for good agronomic performance in two separate SRT planted in two locations. In the old system, the seedling plant was used for two purposes: as a source of roots for pVAC quantification and the stems were used as a source of planting material. In the new system, these functions are performed by different plants. Quantification of pVAC is done only during the dry season, thus avoiding the variation due to changes in DMC that is somewhat related to the timing of harvest. Harvesting of the stems to be used as a source of planting material takes place only when the rains have arrived in the target environment. The planting material does not need to be stored for (sometimes) a long period of time, waiting for the rains to arrive.

8. The transgenic approach

As already demonstrated in this chapter, conventional breeding has been very effective in increasing carotenoids content in the storage roots of cassava. However, breeding cassava is time-consuming and cumbersome due to its heterozygous nature. Adoption of new, improved varieties where cassava plays a key food security role is often low. Farmers tend to be reluctant to shift away from the varieties they have grown for decades [6, 87]. The alternative of turning farmers' preferred varieties into vehicles for delivering pVAC through genetic transformation is very appealing. The technology could deliver exactly the same variety that farmers have grown for years but with increased nutritional value. This is a product that conventional breeding could not offer. Therefore, biotechnology tools have also been included among the strategies that have been considered to deliver biofortified cassava to farmers.

The accumulation of carotenoids in the root involves several genes that, as described above, may have anabolic or catabolic function. This further complicates the conventional breeding strategy, since putting all the desirable alleles of the relevant genes into a single variety are complex and takes time. This would be particularly true if the objective is to silence a gene whose function is to catabolize carotenoids, for example, into ABA (**Figure 3**). Not only enzymes that are directly involved in the making or degrading of carotenoids should be the target of breeding. It has been recently shown that the ORANGE (OR) protein is a posttranscriptional regulator of phytoene synthases (PSYs) in plants [88], which adds another level of intervention (conventional or transgenic) for enhancing carotenoids in cassava roots.

Genetic transformation is not only a viable approach to produce cassava clones with increased pVAC in the roots, but it is also a powerful tool for understanding the individual impact of different genes and alleles in carotenogenesis. Examples of the importance of polymorphisms of single nucleotides—SNPs—from relevant genes such as phytoene synthase have been published [34, 63, 89]. The different studies have demonstrated that the substrates necessary for the activity of relevant enzymes were present in roots of most genotypes. Without this precedent, it would be more difficult to design a genetic modification strategy to increase pVAC contents by inserting new gene combinations in commercial varieties, well established in the markets and accepted by consumers.

As stated above, genetic transformation would add nutritional value to farmers' preferred varieties. In addition, this approach would reduce the time required for developing new varieties since it implies handling few genes of the carotenoid synthetic pathway alone, not an entire genome. Finally, it could ensure that the transgenic variety produces a minimum pVAC as has been the case for genetically modified potato or the Golden Rice [90, 91].

There are already at least five examples of crops in which pVAC contents have been substantially increased using transgenes from the pathway of carotene synthesis: rice, maize, potato, tomato, and canola. They have been guided by promoters that express these genes in specific organs, or constitutively. Genetic transformation of rice, with genes from the carotene pathway [92, 93] has shown that it was possible to increase the TCC in the grain up to 27 times (maximum 37 μg/g, DW), of which more than 80% (>30 μg/g, DW) corresponded to β-carotene. In canola, the β-carotene increase was 50-fold [94]. It has been shown that β-carotene in the potato tuber can be increased >3600 times, reaching 47 μg/g DW, with which 250 g of potato satisfy half the RDA [90].

Both the rice grain and the potato tuber have complemented the carotene synthesis route, which in the case of rice was not very active, with encouraging results suggesting that a similar strategy could be attempted for cassava. In the case of maize, genes of the carotenoids synthesis pathway have been introduced in different combinations producing increases β-carotene and other carotenoids, including complex mixtures of hydroxycarotenoids and ketocarotenoids [95, 96].

The synthesis and accumulation of carotenes in plant storage roots such as cassava and sweet potato are just beginning to be understood at the molecular level. It is not yet very clear how the promoters of carotene synthesis genes are regulated in roots. On the other hand, the perception of foods derived from transgenic crops is not yet totally favorable, with exceptions, although they are safe for human consumption [97]. This forces us to think of strategies to reduce opposition to acceptance, such as replacing bacterial genes, which are used today to genetically modify crops, by plant genes. Genome edition using CRISPR/Cas9 or similar molecular scissors is an alternative for modifying alleles in cassava [62]. However, there are at least three factors that must be taken into account when using genetic modification to increase carotene content in cassava roots: the genes themselves (their coding sequences and their origins), controlling sequences for transcriptional control, and interacting proteins such as PSY and OR for posttranslational control of carotenoid production in plants.

As an example, in the case of cassava, having genetically transformed a wild-type genotype that produces white roots with combinations of the bacterial versions of key carotenoid biosynthetic genes (crtB, crtY, and crtI) significantly increased pVAC levels. However, transgenesis could not exceed the maximum pVAC levels attained by conventional breeding (about 90 μg/g DW in the 2016 harvest, unpublished data). These experiments, however, demonstrated that there were bottlenecks for the synthesis of carotenoids in roots with white parenchyma, such as the absence of a PSY capable of effectively synthesizing phytoene, to keep the route operating so that enough carotenoids were produced and accumulated in the root until the harvest time (usually 11–12 MAP). This bottleneck was solved with the introduction of the CRTB (the bacterial version of PSY in plants) enzyme alone. This modification was responsible for increasing TCC from 0.4 to 22 μg/g (DW) and turning roots from white to yellow [34]. In addition, in the same work, it was confirmed that the PSY enzyme was a limiting step of the pathway, demonstrating that a single SNP in the coding region of the PSY gene could partially explain the difference between white and yellow roots. Thus, deficiency in the carotenoid synthesis pathway in the white cassava root was complemented by providing enzymes more effective than the endogenous ones, possibly not regulated by the plant [34, 62], and by showing allele diversity correlated with better efficiency among PSY endogenous enzymes.

Compared with the increases in TCC obtained in potato by transgenesis (maximum 114.4 μg/g DW; [90]), the 13–31 μg/g (DW) obtained in cassava [34, 63, 98, 99], can be considered only moderate. However, these comparisons are made on a dry weight basis that would favor potato (**Figure 5**) because of its considerable lower DMC and starch contents compared with cassava. The same three genes (crtB, crtY, and crtI) were used in cassava and potato, under the control of the same patatin promoters (in fact it was the same construct). However, the TCC baseline of the nontransgenic control in potato was 5.8 μg/g DW, while for cassava, it was only 1.1 μg/g DW, which improved the chances of increasing TCC in the former. If we accept that the genetic modification of cassava still has a potential to raise the carotenoid content in the root, at levels similar to or higher than those reached in potato, the results with the tuber would indicate the way forward with cassava: raise TCC would be more effective by transgenesis if yellow-rooted (nonwhite) plants were modified.

9. Retention and bioavailability studies

The impact of biofortified cassava roots in the reduction of VAD depends on two factors: (a) how much of the carotenoids present in the raw root at harvest time reach the individual consuming the cassava product; and (b) how much retinol can the individual make with the consumed cassava product. This section describes information regarding these two key steps.

As stated in the introduction, cassava roots have a short shelf life due to PPD. Roots therefore need to be consumed or processed 1–3 days after harvest. A diversity of processing methods has been developed by different cultures resulting in many ethnic products. The easiest and most direct way to consume cassava roots is boiling or steaming them. However, this approach does not allow storing the roots and other processing methods that allow storage for long periods have been developed. Roots can be dried and ground into flour. Drying can be done in an oven, in open air under shaded conditions or by sun drying. Two popular processing methods in West Africa are fufu and gari [100].

A comprehensive review of true retention of carotenoids in cassava roots after alternative processing methods and different storage period has been published [101]. Different authors have highlighted that retention of carotenoids is not only affected by the processing method and length of storage period but also by the cassava genotype [102–104]. Boiling the root is a very simple and popular processing method in many regions of the world, which results in relatively high retentions. There is large variation reported depending on genotypes. Average retention after boiling ranged from 68 [105], around 70 [104], and 74% [100]. DMC in the roots influences true retention of carotenoids after boiling [75, 106]. When the effect of DMC prior to boiling is taken into consideration, retention of carotenoids after this processing method was relatively high (87%) and uniform (retention ranged from 76 to 97%) [75].

Drying the roots soon after harvest is an important method to prevent PPD. Significant differences were observed regarding the drying method employed. The highest β-carotene retention was obtained by oven drying (72%), followed by shade drying (59%), and sun drying (38%) [102]. Similar conclusions were obtained by other authors but with some variation in the average retentions that may be due to the genotype effect [43, 101, 104, 107].

Retention of carotenoids after gari preparation varies widely. Final carotenoids content in gari is a function of the intensity and duration of roasting as well as the duration of the fermentation prior to roasting [101]. Retention reported by Failla and collaborators in 2012 was about 66, 62, and 29%, respectively, for boiling, fufu, and gari [63]. On the other hand, a different study reported average retentions of 74% after boiling, 41 and 22% in raw and cooked fufu, and about 45% in gari [100].

Carotenoids are also lost during storage of the processed product. Chávez and co-workers measured the β-carotene retention during a 4-week storage period with flour and chips that had been either oven-dried or sun-dried [102]. Oven-dried cassava initially retained 72% of β-carotene, which decreased to 40 and 32% after 2 and 4 weeks of storage at room temperature, respectively. Sun-dried cassava had a lower initial β-carotene retention (38%), which was further decreased to 24 and 18% after 2 and 4 weeks of storage, respectively. Retention values for cassava chips were very similar to those of cassava flour. Lower retention levels after storage (at room temperature and in the absence of light) have been reported [108]. Carotenoid content of gari products decreased markedly with time and temperature [109].

The absorption, bioavailability, and conversion of β-carotene, the most prominent carotenoid in yellow cassava, into retinol in the human body depend on several factors. These factors are food- and host-related and depended on, for example, host genotype, availability of fat in the diet, and the food matrix [110]. Beta-carotene in cassava is stored in parenchyma cells, which are more easily destructed in the gastrointestinal tract than, for example, chloroplasts membranes, the primary location of β-carotene in green leafy vegetables. Little is known on the histology of carotenoids stored in cassava roots because of the density of amyloplast. An animal study with gerbils showed a bioconversion factor of 3.7 μg β-carotene to 1 μg of retinol [111]. In a study with 10 healthy American adults, this conversion factor was estimated to be 4.2 μg, but the individual variation between the humans was high (range 0.3–10.6) [112]. A randomized-controlled efficacy study with Kenyan primary school children showed a significant increase of both retinol and β-carotene in the blood after 4 months of feeding with boiled yellow cassava as compared to boiled white cassava [113]. In conclusion, we can say that the β-carotene from cassava is absorbable, bioavailable and converted into retinol in the human body. More research is currently being conducted to provide evidence on the effect of processing in different recipes, as well as for different age groups countries.

10. Conclusions and perspectives

Most of the information described in this chapter relates to the work coordinated and financed by the HarvestPlus initiative. It all began about two decades ago with the simple, yet relevant, idea that the nutritional value of crops could be improved. Since then, the significant progress achieved in different crops was highlighted with the 2016 Food Prize Award to a group of researchers lead by Dr. Howard Bouis.

The significant gains in carotenoids content in cassava roots could only be obtained with the concerted effort of many researchers working in a wide range of disciplines. The first group of cassava varieties with increased levels of carotenoids has been already released in Brazil and in Africa. A second generation of new varieties with higher levels of pVAC and better agronomic performance will soon follow. In addition to the strategic relevance of the germplasm generated, valuable information has been generated ranging from the relationship between carotenoids and DMC in the roots to retention and bioavailability information. Enhanced levels of carotenoids resulted in an unexpected reduction of PPD. The new protocol to screen for carotenoids content based on NIR was developed and prompted further changes in the breeding approach. The different constructs for genetic transformation not only resulted in varying degrees of success but also exposed unexpected responses from the plant, such as the reduction in DMC.

The release of varieties, rich in pVAC, requires strategies for the efficient production of planting material to be distributed to farmers as well as participatory approaches to promote their adoption. Deploying these varieties will provide excellent opportunities for nutritional studies and development of new food products and alternative processing methods. The concerted effort of many researchers and institutions and the valuable financial support of key donors and investors have been motivated by the magnitude of VAD. Hopefully, the scientific community will soon be able to document the impact of these efforts in the livelihood of millions of people affected by VAD.

Author details

Hernán Ceballos[1,2]*, Fabrice Davrieux[3], Elise F. Talsma[2], John Belalcazar[1], Paul Chavarriaga[1] and Meike S. Andersson[2]

*Address all correspondence to: h.ceballos@cgiar.org

1 International Center for Tropical Agriculture (CIAT), Cali, Colombia

2 HarvestPlus Organization, Cali, Colombia

3 Centre de Cooperation Internationale en Recherche Agronomique pour le Developpement (CIRAD), UMR Qualisud, St Pierre, Reunion Island, France

References

[1] Babu L, Chatterjee SR. Protein content and amino acid composition of cassava tubers and leaves. Journal of Root Crops. 1999;**25**(20):163-168

[2] Balagopalan C. Cassava utilization in food, feed and industry. In: Hillocks RJ, Tresh JM, Bellotti AC, editors. Cassava: biology, production and utilization. Wallingford, UK and New York, USA: CABI Publishing; 2002. pp. 301-318

[3] Caccamisi DS. Cassava: Global production and market trends. Chronica Horticulturae. 2010;**50**(2):15-18

[4] Nweke F. New Challenges in the Cassava Transformation in Nigeria and Ghana. EPT Discussion Paper No. 118. Washington DC, USA: International Food Policy Research Institute (IFPRI); 2004

[5] Sasson A. Food security for Africa: An urgent global challenge. Agriculture & Food Security. 2012;**1**:1-16

[6] Tarawali G, Iyangbe C, Udensi UE, Ilona P, Osun T, Okater C, Asumugha GN. Commercial-scale adoption of improved cassava varieties: A baseline study to highlight constraints of large-scale cassava based agro-processing industries in Southern Nigeria. Journal of Food, Agriculture and Environment. 2012;**10**:689-694

[7] Burns A, Gleadow R, Cliff J, Zacarias A, Cavagnaro T. Cassava: The drought, war, and famine crop in a changing world. Sustainability. 2010;**2**:3572-35607. doi:10.3390/su2113572

[8] Dixon AGO, Bandyopadhyay R, Coyne D, Ferguson M, Ferris RSB, Hanna R, Hughes J, Ingelbrecht I, Legg J, Mahungu N, Manyong V, Mowbray D, Neuenschwander P, Whyte J, Hartmann P, Ortiz R. Cassava: From poor Farmers' crop to pacesetter of African rural development. Chronica Horticulturae. 2003;**43**:8-15

[9] Jarvis A, Ramirez-Villegas J, Campo BVH, Navarro-Racines C. Is cassava the answer to African climate change adaptation? Tropical Plant Biology 2012;**5**:9-29

[10] OECD-Organisation for Economic Co-operation and Development. Consensus Document on Compositional Considerations for New varieties of Cassava (*Manihot esculenta* Crantz): Key Food and Feed Nutrients, Anti-nutrients, Toxicans and Allergens. Paris: OECD Environment, Health and Safety Publications, Series on the Safety of Novel Foods and Feeds No. 18; 2009. 50 p. (http://www.oecd.org/env/ehs/biotrack/46815306.pdf)

[11] Prakash A. Cassava: International market Profile. Background Paper for the Competitive Commercial Agriculture in Sub–Saharan Africa (CCAA) Study. Trade and Markets Division. Rome, Italy: Food and Agriculture Organisation of the United Nations; 2008

[12] Stapleton G. Global starch market outlook and competing starch raw materials for starches by product segment and region. In: Cassava Starch World 2012. Phnom Penh, Cambodia: Centre for Management Technology (CMT); February, 2012

[13] Liang H, Ren J, Gao Z, Gao S, Luo X, Dong L, Scipioni A. Identification of critical success factors for sustainable development of biofuel industry in China based on grey decision-making trial and evaluation laboratory (DEMATEL). Journal of Cleaner Production. 2016;**131**:500-508

[14] Sriroth K, Piyachomkwan K, Wanlapatit S, Nivitchanyong S. The promise of a technology revolution in cassava bioethanol: From Thai practice to the world practice. Fuel. 2010;**89**:1333-1338

[15] Sánchez T, Mafla G, Morante N, Ceballos H, Dufour D, Calle F, Moreno X, Pérez JC, Debouck D. Screening of starch quality traits in cassava (*Manihot esculenta* Crantz). Starch-Starke. 2009;**61**:12-19

[16] Combs GF. The Vitamins Fundamental Aspects in Nutrition and Health. London: Academic Press; 1998. pp. 93-138

[17] WHO. Global Prevalence of Vitamin A deficiency in Populations at Risk 1995–2005. Geneva, Switzerland: World Health Organization; 2009

[18] Alós E, Maria Rodrigo J, Zacarias L. Manipulation of carotenoid content in plants to improve human health. Claudia C Stange (ed) Subcellular Biochemistry. Carotenoids in Nature. 2016;**79**:311-343

[19] Bouis HE, Hotz C, McClafferty B, Meenakshi JV, Pfeiffer WH. Biofortification: A new tool to reduce micronutrient malnutrition. Food and Nutrition Bulletin. 2011;**32**:S31-S40

[20] West KP. Vitamin A deficiency disorders in children and women. Food and Nutrition Bulletin. 2003;**24**:S78-S90

[21] Hamer DH, Keusch GT. Vitamin A deficiency: slow progress towards elimination. The Lancet. 2015;**3**:502-503

[22] Dwivedi SL, Sahrawat KL, Rai KN, Blair MW, Andersson MS, Pfeiffer W. Nutritionally enhanced staple food crops. Plant Breeding Reviews. 2012;**36**:173-293

[23] Montagnac JA, Davis CR, Tanumihardjo SA. Nutritional value of cassava for use as a staple food and recent advances for improvement. Comprehensive Reviews in Food Science and Food Safety. 2009;**18**:181-194. doi:10.1111/j.1541-4337.2009.00077

[24] Botella-Pavia P, Rodriguez-Concepcion M. Carotenoid biotechnology in plants for nutritionally improved foods. Physiologia Plantarum 2006;**126**:369-381

[25] Demmig-Adams B, Adams WW 3rd. Antioxidants in photosynthesis and human nutrition. Science. 2002;**298**(5601):2149-2153

[26] Chávez AL, Sánchez T, Jaramillo G, Bedoya JM, Echeverry J, Bolaños EA, Ceballos H, Iglesias CA. Variation of quality traits in cassava roots evaluated in landraces and improved clones. Euphytica. 2005;**143**:125-133

[27] Posada CA, López-G A, Ceballos H. Influencia de harinas de yuca y de batata sobre pigmentación, contenido de carotenoides en la yema y desempeño productivo de aves en postura. Acta Agronómica. 2006;**55**(3):47-54

[28] USDA. National Nutrient Database for Standard Reference, Release 28. Washington DC, USA: Agricultural Research Service, US Department of Agriculture; 2016

[29] Gegios A, Amthor R, Maziya-Dixon B, Egesi C, Mallowa S, Nungo R, Gichuki S, Mbanaso A, Manary MJ. Children consuming cassava as a staple food are at risk for inadequate zinc, iron, and vitamin A intake. Plant Foods for Human Nutrition 2010;**65**:64-70

[30] Alves AAC. Cassava botany and physiology. In: Hillocks RJ, Thresh JM, Bellotti A (eds). Cassava: Biology, Production and Utilization. Wallingford, UK: CABI Publishing; 2002. pp. 67-89

[31] Nzwalo H, Cliff J. Konzo: from poverty, cassava, and cyanogen intake to toxico-nutritional neurological disease. PLoS Neglected Tropical Diseases. 2011;**5**(6):e1051. doi:10.1371/journal.pntd.0001051

[32] Vandegeer R, Miller RE, Bain M, Gleadow RM, Cavagnaro TR. Drought adversely affects tuber development and nutritional quality of the staple crop cassava (*Manihot esculenta* Crantz). Functional Plant Biology. 2013;**40**:195-200

[33] Stange C, Fuentes P, Handford M, Pizarro L. *Daucus carota* as a novel model to evaluate the effect of light on carotenogenic gene expression. Biological Research. 2008;**41**:289-301

[34] Welsch R, Arango J, Bär C, Salazar B, Al-Babili S, Beltrán J, Chavarriaga P, Ceballos H, Tohme J, Beyer P. Provitamin A accumulation in cassava (*Manihot esculenta*) roots driven by a single nucleotide polymorphism in a phytoene synthase gene. Plant Cell. 2010;**22**:3348-3356. DOI: 10.1105/tpc.110.077560

[35] Chávez AL, Bedoya JM, Sánchez T, Iglesias C, Ceballos H, Roca W. Iron, carotene, and ascorbic acid in cassava roots and leaves. Food and Nutrition Bulletin. 2000;**21**:410-413

[36] Ceballos H, Chávez AL, Sánchez T, Bedoya JM, Echeverri J, Tohme J. Genetic potential to improve carotene content of cassava and strategies for its deployment. The Journal of Nutrition. 2002;**132**(9S):2988S

[37] Maziya-Dixon B, Kling JB, Menkir A, Dixon A. Genetic variation in total carotene, iron, and zinc contents of maize and cassava genotypes. Food and Nutrition Bulletin. 2000;**21**(4):410-413

[38] Maravalhas N. Carotenoides de *Manihot esculenta* Crantz. Manaus, Brazil: Publicacao No. 6, Institute Nacional de Pesquisas de Amazonica; 1961, pp. 33-38 (in Portuguese)

[39] Guimaraes ML, Barros MSCD. The Occurrence of β-carotene in Varieties of Yellow Cassava. Rio de Janeiro, Brazil: Boletim Technologia Agricolae Alimentar. Ministereo de Agricultura; 1971, 4 p. (in Portuguese)

[40] Arkcoll DB. Some interesting cassava varieties from the Amazon region. Acta Amazonica 1981;**11**:207-211 (in Portuguese)

[41] Moorthy SN, Jos JS, Nair RB, Sreekumari MT. Variability of B-carotene content in cassava germplasm. Food Chemistry. 1990;**36**:233-236

[42] Jos JS, Nair SG, Moorthy SN, Nair RB. Carotene enhancement in cassava. Journal Root of Crops. 1990;**16**:5-11

[43] Iglesias C, Mayer J, Chávez AL, Calle F. Genetic potential and stability of carotene content in cassava roots. Euphytica. 1997;**94**:367-373

[44] Oduro KA. Some characteristics of yellow pigmented cassava. In: Terry ER, Oduro KA, Caveness F, editors. Tropical Roots Crops Research Strategies for the 1980s. Ottawa, Canada: IDRC; 1981.

[45] Ceballos H, Morante M, Sánchez T, Ortiz D, Aragón I, Chávez AL, Pizarro M, Calle F, Dufour D. Rapid cycling recurrent selection for increased carotenoids content in cassava roots. Crop Science 2013;**53**:2342-2351

[46] Sánchez T, Chávez AL, Ceballos H, Rodriguez-Amaya DB, Nestel P, Ishitani M. Reduction or delay of post-harvest physiological deterioration in cassava roots with higher carotenoid content. Journal of the Science of Food and Agriculture. 2006;**86**(4):634-639

[47] Morante N, Sánchez T, Ceballos H, Calle F, Pérez JC, Egesi C, Cuambe CE, Escobar AF, Ortiz D, Chávez AL. Tolerance to post-harvest physiological deterioration in cassava roots. Crop Science 2010;**50**:1333-1338

[48] Bouvier F, Isner J-C, Dogbo O, Camara B. Oxidative tailoring of carotenoids: A prospect towards novel functions in plants. Trends in Plant Science. 2005;**10**(4):187-194

[49] Yeum K-J, Russell RM. Carotenoid bioavailability and bioconversion. Annual Review of Nutrition 2002;**22**:483-504

[50] Maziya-Dixon BB, Dixon AGO. Carotenoids content of yellow-fleshed cassava genotypes grown in four agroecological zones in Nigeria and their Retinol Activity Equivalents (RAE). Journal of Food, Agriculture & Environment. 2015;**13**(2):63-69

[51] Hershey CH, Ocampo-N CH. New marker genes found in cassava. Cassava Newsletter. 1989;**13**(1):1-5

[52] Akinwale MG, Aladesanmwa RD, Akinyele BO, Dixon AGO, Odiyi AC. Inheritance of β-carotene in cassava (*Manihot esculenta* Crantz). International Journal of Genetics and Molecular Biology 2010;**2**:198-201

[53] Morillo-Coronado, Y. Herencia del contenido de carotenos en raíces de yuca (*Manihot esculenta* Crantz). Ph.D. Dissertation. Universidad Nacional de Colombia Palmira Campus; 2009

[54] Morillo-C Y, Sánchez T, Morante N, Chávez AL, Morillo-C AC, Bolaños A, Ceballos H. Estudio preliminar de herencia del contenido de carotenoides en raíces de poblaciones segregantes de yuca (*Manihot esculenta* Crantz). Acta Agronómica. 2012;**61**(3):253-264

[55] Njoku DN, Gracen VE, Offei SK, Asante IK, Egesi CN, Kulakow P, Ceballos H. Parent-offspring regression analysis for total carotenoids and some agronomic traits in cassava. Euphytica. 2015;**206**:657-666

[56] Esuma W, Kawuki RS, Herselman L, Labuschagne MT. Stability and genotype by environment interaction of provitamin A carotenoid and dry matter content in cassava in Uganda. Breeding Science 2016;**66**:434-443

[57] Maroya NG, Kulakow P, Dixon AGO, Maziya-Dixon B, Bakare MA. Genotype × environment interaction of carotene content of yellow-fleshed cassava genotypes in Nigeria. Journal of Life Sciences 2012;**6**:595-601

[58] Ssemakula G, Dixon A. Genotype X environment interaction, stability and agronomic performance of carotenoid-rich cassava clones. Scientific Research and Essay. 2007;**2**(9):390-399

[59] Njenga P, Edema R, Kamau J. Combining ability for beta-carotene and important quantitative traits in a cassava F1 population. Journal of Plant Breeding and Crop Science. 2014;**6**(2):24-30

[60] Hirschberg J. Carotenoid biosynthesis in flowering plants. Current Opinion in Plant Biology 2001;**4**:210-218

[61] Arango J, Wüst F, Beyer P, Welsch R. Characterization of phytoene synthases from cassava and their involvement in abiotic stress-mediated responses. Planta 2010;**232**:1251-1262

[62] Chavarriaga-Aguirre P, Brand A, Medina A, Prías M, Escobar R, Martinez J, Díaz P, López C, Roca WR, Tohme J. The potential of using biotechnology to improve cassava: a review. In Vitro Cellular & Developmental Biology. Plant. 2016;**52**:461-478

[63] Failla ML, Chitchumroonchokchai C, Siritunga D, De Moura FF,Fregene M, Sayre RT. Retention during processing and bioaccessibility of β-Carotene in high β-Carotene transgenic cassava root. Journal of Agricultural and Food Chemistry. 2012. DOI: 10.1021/jf204958w

[64] Diretto G, Welsch R, Tavazza R, Mourgues F, Pizzichini D, Beyer P, Giuliano G. Silencing of β-carotene hydroxylase increases total carotenoid and β-carotene levels in potato tubers. BMC Plant Biology 2007;**7**:11. DOI: 10.1186/1471-2229-7-11

[65] Kim SH, Ahn YO, Ahn MJ, Lee HS, Kwak SS. Down-regulation of β -carotene hydroxylase increases β-carotene and total carotenoids enhancing salt stress tolerance in transgenic cultured cells of sweetpotato. Phytochemistry 2012;**74**:69-78

[66] Esuma W, Rubaihayo P, Pariyo A, Kawuki R, Wanjala B, Nzuki I, Harvey JJW, Baguma Y. Genetic diversity of provitamin A cassava in Uganda. Journal of Plant Studies 2012;**1**(1):60-71

[67] Esuma W, Herselman L, Labuschagne MT, Ramu P, Lu F, Baguma Y, Buckler ES, Kawuki RS. Genome-wide association mapping of provitamin A carotenoid content in cassava. Euphytica 2016;**212**:97-110

[68] Fortes Ferreira C, Alves E, Nogueira Pestana K, Theodoro Junghans D, Kobayashi AK, Santos VDJ, Pereira Silva R, Henrique Silva P, Soares E, Fukuda W. Molecular characterization of Cassava (*Manihot esculenta* Crantz) with yellow-orange roots for beta-carotene improvement. Crop Breeding and Applied Biotechnology. 2008;**8**:23-29

[69] Morillo-C AC, Morillo-C Y, Fregene M, Ramirez H, Chávez AL, Sánchez T, Morante N, Ceballos LH. Diversidad genética y contenido de carotenos totales en accesiones del germoplasma de yuca (*Manihot esculenta* Crantz). Acta Agronómica. 2011;**60**(2):97-107

[70] Njoku DN, Gracen VE, Offei SK, Asante IK, Danquah EY, Egesi CN, Okogbenin E. Molecular marker analysis of F1 progenies and their parents for carotenoids inheritance in African cassava (*Manihot esculenta* Crantz). African Journal of Biotechnology. 2014;13:3999-4007

[71] Kim JE, Rensing KH, Douglas CJ, Cheng KM. Chromoplasts ultrastructure and estimated carotene content in root secondary phloem of different carrot varieties. Planta 2010;**231**:549-558

[72] Alves AAC, Setter TL. Response of cassava to water deficit: leaf area growth and abscisic acid. Crop Science 1999;**40**:131-137

[73] Chávez AL, Ceballos H, Rodriguez-Amaya DB, Pérez JC, Sánchez T, Calle F, Morante N. Sampling variation for carotenoids and dry matter contents in cassava roots. Journal of Root Crops. 2008;**34**(1):43-49

[74] Ortiz D, Sánchez T, Morante N, Ceballos H, Pachón H, Duque MC, Chávez AL, Escobar AF. Sampling strategies for proper quantification of carotenoids content in cassava breeding. Journal of Plant Breeding and Crop Science. 2011;**3**(1):14-23

[75] Ceballos H, Luna J, Escobar AF, Pérez JC, Ortiz D, Sánchez T, Pachón H, Dufour D. Spatial distribution of dry matter in yellow fleshed cassava roots and its influence on carotenoids retention upon boiling. Food Research International. 2012;**45**:52-59

[76] Rodriguez-Amaya DB. A Guide to Carotenoid Analysis in Foods. Washington DC: ILSI Press; 2001

[77] Rodriguez-Amaya DB, Kimura M. HarvestPlus handbook for carotenoid analysis. HarvestPlus Technical Monograph 2. Washington, DC and Cali, Colombia: International Food Policy Research Institute (IFPRI) and International Center for Tropical Agriculture (CIAT); 2004

[78] Ceballos H, Hershey C and Becerra-López-Lavalle LA. New approaches to cassava breeding. Plant Breeding Reviews. 2012;**36**:427-504

[79] Osborne BG, Fearn T, Hindle PH. Practical NIR Spectroscopy with Applications in Food and Beverage Analysis. Longman Scientific and Technical; Harlow UK. 1993

[80] Pasquini C. Near infrared spectroscopy: Fundamentals, practical aspects and analytical applications. Journal of the Brazilian Chemical Society. 2003;**14**:198-219

[81] McGoverin CM, Weeranantanaphan J, Downey G, Manley M. The application of near infrared spectroscopy to the measurement of bioactive compounds in food commodities. Journal of Near Infrared Spectroscopy. 2010:**18**:87-111

[82] Sánchez T, Ceballos H, Dufour D, Ortiz D, Morante N, Calle F, Zum Felde T, Davrieux F. Carotenoids and dry matter prediction by NIRS and hunter color in fresh cassava roots. Food Chemistry. 2014;**151**:444-451

[83] Shenk JS, Westerhaus, MO, Berzaghi P. Investigation of a LOCAL calibration procedure for near infrared instruments. Journal of Near Infrared Spectroscopy. 1997;**5**:223-232

[84] Davrieux F, Dufour D, Dardenne P, Belalcazar J, Pizarro M, Luna J, Londoño L, Jaramillo A, Sanchez T, Morante N, Calle F, Becerra LA, Ceballos H. LOCAL regression algorithm improves NIRS predictions when the target constituent evolves in breeding populations. JNIRS. 2016;**24**:109-117

[85] Belalcazar J, Dufour D, Andersson MS, Pizarro M, Luna J, Londoño L, Morante N, Calle F, Jaramillo AM, Pino L, Becerra López-Lavalle LA, Davrieux F and Ceballos H. High-throughput phenotyping and improvement in breeding cassava for increased carotenoids in the roots. Crop Science. 2016;**56**:2916-2925

[86] HarvestPlust Biofortification Progress Briefs; 2014. Available online at: www.HarvestPlus.org

[87] Manu-Aduening JA, Lamboll RI, Ampong Mensah G, Lamptey JN, Moses E, Dankyi AA, Gibson RW. Development of superior cassava cultivars in Ghana by farmers and scientists: The process adopted, outcomes and contributions and changed roles of different stakeholders. Euphytica. 2006;**150**:47-61

[88] Zhou X, Welsch R, Yan Y, Álvarez D, Riediger M, Yuana H, Fish T, Liu, Thannhauser TW, Li L. *Arabidopsis* OR proteins are the major posttranscriptional regulators of phytoene synthase in controlling carotenoid biosynthesis. Proceedings of the National Academy of Science. 2015;**112**(11):3558-3563

[89] Arango-Mejia J. Análisis de la expresión de los genes de la ruta biosintética de carotenos y cuantificación de carotenos en hojas y raíces de plantas de yuca de diferentes edades. Bogotá, Colombia: Trabajo de grado. Pontificia Universidad Javeriana, Facultad de Ciencias, Carrera de Biología; 2006. 75P

[90] Diretto G, Al-Babili S, Tavazza R, Papacchioli V,Beyer P, Giuliano G metabolic engineering of potato carotenoid content through tuber-specific overexpression of a bacterial mini-pathway. PLoS One. 2007;**2**(4):e350

[91] Tang G, Qin J, Dolnikowski GG, Russell RM, Grusak MA. Golden Rice is an effective source of vitamin A. American Journal of Clinical Nutrition. 2009;**89**(6):1776-1783

[92] Paine JA, Shipton CA, Chaggar S, Howells RM, Kennedy MJ, Vernon G, Wright SY, Hinchliffe E, Adams JL, Silverstone AL, Drake R. Improving the nutritional value of Golden Rice through increased pro-vitamin A content. Nature Biotechnology. 2005;**23**: 482-487

[93] Ye X, Al Babili S, Kloti A, Zhang J, Lucca P, Beyer P, Potrykus I. Engineering pro-vitamin A (β-carotene) biosynthesis pathway into (carotenoid free) rice endosperm. Science 2000;**287**:303-305

[94] Shewmaker CK, Sheehy JA, Daley M, Colburn S, Ke DY. Seed-specific over expression of phytoene synthase: increase in carotenoids and other metabolic effects. The Plant Journal. 1999;**20**(4):401-412.

[95] Zhu C, Naqvi S, Breitenback J, Sandmann G, Christou P, Capell T. Combinatorial genetic transformation generates a library of metabolic phenotypes for the carotenoid pathway in maize. Proceedings of the National Academy of Science. 2008;**105**(47):18232-18237

[96] Naqvi S, Zhu C, Farre G, Ramessar K, Bassie L, Breitenbach J, Perez D, Ros G, Sandmann G, Capell T, Christou P. Transgenic multivitamin corn through biofortification of endosperm with three vitamins representing three different metabolic pathways. Proceedings of the National Academy of Science. 2009;**106**(19):7762-7767

[97] Gonzalez C, Johnson N, Qaim M. Consumer acceptance of second-generation GM-food: The case of bio-fortified cassava in the north-east of Brazil. Journal of Agricultural Economics. 2009;**60**(3):304-324

[98] Bonilla A. Análisis de expresión de una miniruta de síntesis de carotenos en yuca (*Manihot esculenta* Crantz) transgénica por medio de qRT-PCR (reacción en cadena de la polimerasa en tiempo real). Tesis: Universidad del Quindío, Programa de Biología; 2010

[99] Chavarriaga P. Biotecnología y mejoramiento convencional para incrementar el valor nutricional de la raíz de yuca (*Manihot esculenta* Crantz). PhD thesis. Cali, Colombia: Facultad de Ciencias Naturales y Exactas, Universidad del Valle; 2013

[100] Maziya-Dixon B, Awoyale W, Dixon A. Effect of processing on the retention of total carotenoid, iron and zinc contents of yellow-fleshed cassava roots. Journal of Food and Nutrition Research. 2015;**3**:483-488

[101] De Moura FF, Miloff A, Boy E. Retention of provitamin A carotenoids in staple crops targeted for biofortification in Africa: cassava, maize and sweet potato. Critical Reviews in Food Science and Nutrition 2015;**55**:1246-1269

[102] Chávez AL, Sánchez T, Ceballos H, Rodriguez-Amaya DB, Nestel P, Tohme J, Ishitani M. Retention of carotenoids in cassava roots submitted to different processing methods. Journal of the Science of Food and Agriculture 2007;**87**:388-393

[103] Carvalho LMJ, Oliveira ARG, Godoy RLO, Pacheco S, Nutti MR, de Carvalho JLV, Pereira EJ, Fukuda WG. Retention of total carotenoid and β-carotene in yellow sweet cassava (*Manihot esculenta* Crantz) after domestic cooking. Food & Nutrition Research 2012;**56**:15788. DOI: 10.3402/fnr.v56i0.15788

[104] Vimala B, Thushara R, Nambisan B, Sreekumar J. Effect of processing on the retention of carotenoids in yellow-fleshed cassava (*Manihot esculenta* Crantz) roots. International Journal of Food Science and Technology 2011;**46**:166-169

[105] Berni P, Chitchumroonchokchai C, Canniatti-Brazaca SG, De Moura FF, Mark L, Failla M. Impact of genotype and cooking style on the content, retention, and bioacessibility of β-carotene in biofortified cassava (*Manihot esculenta* Crantz) conventionally bred in Brazil. Journal of Agricultural and Food Chemistry 2014;**62**:6677-6686

[106] Thakkar SK, Huo T, Maziya-Dixon B, Failla ML. Impact of style of processing on retention and bioaccessibility of β-Carotene in cassava (*Manihot esculenta* Crantz). Journal of Agricultural and Food Chemistry 2009;**54**:1344-1348

[107] Maziya-Dixon B, Dixon AGO, Ssemakula, G. Changes in total carotenoid content at different stages of traditional processing of yellow-fleshed cassava genotypes. International Journal of Food Science and Technology 2009;**44**:2350-2357

[108] Oliveira RGA, de Carvalho MJL, Nutti RM, de Carvalho LVJ, Gonçalves Fukuda W. Assessment and degradation study of total carotenoid and β-carotene in bitter yellow cassava (*Manihot esculenta* Crantz) varieties. African Journal of Food Science 2010;**4**:148-155

[109] Bechoff A, Chijioke U, Tomlins KI, Govinden P, Ilona P, Westby A, Boy E. Carotenoid stability during storage of yellow gari made from biofortified cassava or with palm oil. Journal of Food Composition and Analysis 2015;**44**:36-44

[110] Castenmiller JJ, West CE. Bioavailability and bioconversion of carotenoids. Annual Review of Nutrition 1998;**18**:19-38

[111] Howe JA, Maziya-Dixon B, Tanumihardjo SA. Cassava with enhanced β-carotene maintains adequate vitamin A status in Mongolian gerbils (*Meriones unguiculatus*) despite substantial cis-isomer content. British Journal of Nutrition 2009;**102**:342-349

[112] La Frano MR, Woodhouse LR, Burnett DJ, Burri BJ. Biofortified cassava increases β-carotene and vitamin A concentrations in the TAG-rich plasma layer of American women. British Journal of Nutrition 2013;**110**:310-320

[113] Talsma EF, Brouwer ID, Verhoef H, Mbera GN, Mwangi AM, Demir AY, Maziya-Dixon B, Boy E, Zimmermann MB, Melse-Boonstra A. Biofortified yellow cassava and vitamin A status of Kenyan children: a randomized controlled trial. American Journal of Clinical Nutrition 2016;**103**:258-267